MEDICAL REVIEW SERIES

Neuroscience

Notice

Medicine is an ever-changing science. As new research and clinical experience broaden our knowledge, changes in treatment and drug therapy are required. The authors and publisher of this work have checked with sources believed to be reliable in their efforts to provide information that is complete and generally in accord with the standards accepted at the time of publication. However, in view of the possibility of human error or changes in medical sciences, neither the authors nor the publisher nor any other party who has been involved in the preparation or publication of this work warrants that the information contained herein is in every respect accurate or complete, and they are not responsible for any errors or omissions or for the results obtained from use of such information. Readers are encouraged to confirm the information contained herein with other sources. For example and in particular, readers are advised to check the product information sheet included in the package of each drug they plan to administer to be certain that the information contained in this book is accurate and that changes have not been made in the recommended dose or in the contraindications for administration. This recommendation is of particular importance in connection with new or infrequently used drugs.

MEDICAL REVIEW SERIES

Neuroscience

Written by
Luis Yarzagaray, M.D.
Clinical Professor of Neurosurgery
University of Illinois at Chicago
Chicago, IL

Nikos M. Linardakis, M.D.
Editor-in-Chief
Digging Up the Bones Medical Review Series
Chief Medical Officer, Hygeia-Epione Medical Corp.
Salt Lake City, Utah

McGraw-Hill
Health Professions Division

New York St. Louis San Francisco Auckland Bogotá Caracas Lisbon London
Madrid Mexico City Milan Montreal New Delhi San Juan Singapore Sydney
Tokyo Toronto

McGraw-Hill
A Division of The McGraw-Hill Companies

NEUROSCIENCE: Digging Up the Bones

Copyright © 2000 by *The McGraw-Hill Companies, Inc.* All rights reserved. Printed in the United States of America. Except as permitted under the United States Copyright Act of 1976, no part of this publication may be reproduced or distributed in any form or by any means, or stored in a data base or retrieval system, without prior written permission of the publisher.

1234567890 MALMAL 99

ISBN 0-07-038369-3

This book was set in Times Roman by V & M Graphics, Inc.
The editors were John Dolan and Susan R. Noujaim.
The production supervisor was Helene G. Landers.
The cover designer was Mathew Dvorozniak.

This book was printed on acid-free paper.

Library of Congress Cataloging-in-Publication Data

Linardakis, Nikos M.
    Neuroscience / written by Nikos M. Linardakis, Luis Yarzagaray.
       p.   c.m. — (Digging up the bones)
    Includes bibliographical references and index.
    ISBN 0-07-038369-3
    1. Neurosciences Outlines, syllabi, etc.   I. Yarzagaray Luis.—
II. Title.   III. Series: Digging up the bones medical review series.
    [DNLM: 1. Brain. 2. Nervous System.   WL 300 L735n   2000]
   RC343.6.L56   2000
   612.8—DC21
   DNLM/DLC
   for Library of Congress   99-31930
                                                                                        CIP

To My Wife and My Entire Family
for their support, encouragement
and devotion

Luis Yarzagaray

Contents

Preface *ix*

Chapter 1
The Brain — 1

Chapter 2
The Frontal Lobe — 7

Chapter 3
The Parietal Lobe — 15

Chapter 4
Language, the Mechanism of Speech, and Aphasias — 21

Chapter 5
The Occipital Lobe — 25

Chapter 6
The Temporal Lobe — 29

Chapter 7
Insula — 33

Chapter 8
The Limbic System — 35

Chapter 9
The Basal Ganglia — 39

Chapter 10
The Thalamus — 43

Chapter 11
The Hypothalamus and Pituitary Gland — 49

Chapter 12
The Corpus Callosum   55

Chapter 13
The Ventricular System   59

Chapter 14
The MidBrain   65

Chapter 15
The Pons   73

Chapter 16
Syndromes of the Pons   79

Chapter 17
The Medulla Oblongata   83

Chapter 18
The Cerebellum   89

Chapter 19
The Cranial Nerves   97

Chapter 20
The Meninges   123

Chapter 21
The Spinal Cord   129

Chapter 22
The Cervical and Brachial Plexus   141

Chapter 23
The Lumbosacral Plexus   147

Chapter 24
The Autonomic Nervous System   153

Chapter 25
Brain Tumors   159

Bibliography   177

Index   181

Preface

Prepare to learn! In the *Digging Up the Bones*® review series, we will cover several review items that are essential for the United States Medical Licensing Examinations and course exam review. Please be certain to pay close attention to the *italicized* and **bold**faced items as they are likely to show up on your examinations. I believe the following will be an enormous tool in "digging up the bones" of neuroscience and to increase your understanding of the material. I recommend placing your own notes on the side margins, and review this material. I recommend placing your own notes on the side margins, and review this material at least twice to *understand* and MEMORIZE it. The clinical scenarios will help in this memorization process. Overall, the information presented should be of tremendous help.

The volume of the *Digging Up the Bones*® medical review series was created to present neuroscience highlights in a straightforward, easy-to-learn format. Please take the time to learn ALL of the illustrations and noted landmarks.

This review series was created as a concise *summary* of the different diagnoses, treatments, and related facts within medicine. I know you will appreciate this volume's *brief* and *concise* presentation of the facts within neuroscience. Furthermore, this will increase your mastery of the necessary material to help in your medical practice in the future.

Dr. Luis Yarzagaray, Chairman of Neurosurgery with over 30 years experience, presents neuroscience in an effortless fashion. His talents have helped thousands of patients and now his summary review of neuroscience will help in your mastery of this topic.

Digging Up the Bones® Medical Review Series has received recognition from around the world through the shared comments with physicians and students. I know your efforts will pay off, and you will share these skills with others in the years to come.

Now, let's get started!

Nikos M. Linardakis, M.D.

The Brain   1

The brain is that portion of the cerebrospinal system which is contained inside the head. In general, we can say that it has five parts: (1) two cerebral hemispheres, (2) the diencephalon, (3) the cerebral peduncles, (4) the pons cerebellum, and (5) the medulla oblongata.

CEREBRAL HEMISPHERE

The cerebral hemispheres constitute the larger portion of the encephalon. They are located above the tentorium and fill the cranial cavity. The interhemispheric fissure separates these two hemispheres.

Every hemisphere presents an outer surface, which has a convexity shape in the outer part. The medial aspect of the hemispheres corresponds to the homologous medial part, and it is in direct contact with the falx cerebri. The inferior surface, or base, rests over the orbit, middle fossa, and tentorium. The brain has an inner, or subcortical, structure, which is called the white matter; and an outer stratum, which is called the gray substance. If one studies the outer, or convex, portion of the cerebral hemispheres, three large fissures can be seen.

1. The *rolandic fissure* begins above the upper mid-portion of the hemispheres and runs downward and somewhat forward. It divides the outer part of the hemisphere into two almost equal parts. It extends from the superior portion of the brain to the sylvian fissure, also called lateral fissure. Because of the fact that it divides the lateral surface of the hemispheres into two almost equal halves, it is also called the central fissure.
2. The *sylvian fissure*, also called the lateral fissure, is an easy anatomic landmark to identify. It begins at the base of the brain at the sylvian valley, curves outward toward the external surface of the brain, and extends back all the way to the parietal lobe. The upper portion, which limits the great horizontal fissure of sylvius, is the so-called *superior operculum*, which is integrated by the frontal and parietal lobes. The inferior limit, or perhaps called the inferior lip of the sylvian fissure, is

the superior temporal gyrus, which starts at the pole of the temporal lobe and runs all the way back to the junction with the parietal lobe.
3. The *parieto-occipital fissure* is located posteriorly in the brain, and is limited by the parietal and occipital lobes. This is also known as the Simian sulcus because of its relationship to the sulcus found in a monkey.

With the presence of these fissures, the brain can be divided into four lobes.

1. The *frontal lobe*, in front of the central sulcus all the way to the tip of the frontal lobe.
2. The *parietal lobe*, limited by the rolandic fissure in front, the superior border of the brain above, the sylvian fissure below, and the parieto-occipital fissure posteriorly.
3. The *occipital lobe* is located behind the parieto-occipital fissure. This fissure is very poorly demarcated. It begins at the superior border of the hemisphere and runs down to the inferior border of the brain. The occipital lobe is the smallest of the four lobes.
4. The *temporal lobe* has as upper limits the fissure of sylvius and the inferior border of the brain below, and posteriorly, it is limited by the parieto-occipital fissure. Keep in mind that this is describing the *lateral* aspect of the cerebral hemisphere or perhaps, it would be better to say the *lateral* surface of the brain.

MEDIAL SURFACE OF THE CEREBRAL HEMISPHERE

The *medial* surface of the cerebral hemisphere shows a very extensive white structure known as the *corpus callosum*, and that will be subsequently described in detail. In general, one can say that the corpus callosum looks like the inferior extremity; having anteriorly and inferiorly: a leg and a knee, or genu, that has a curvature that extends to the frontal lobe (see Figure 1-1). The genu continues with the *body* of the corpus callosum—which can be thought of as the thigh and extends back toward the occipital lobe. This area becomes very thickened and is called the *splenium*, which, using the analogy of the human leg, would correspond to the buttock. That, of course, is a simple analogy to help the reader to remember the anatomy of the corpus callosum.

The medial aspect of the cerebral hemisphere, in contraposition with the external aspect of the brain, presents only a large sulcus, the *callosomarginal fissure*, which outlines a gyrus known as *cingulate gyrus* in the upper limits. The cingulate gyrus is resting over the corpus callosum, and is separated from it by the *sulcus* of the corpus callosum (the sulcus separates the two). Coming back to the subject of the cingulate sulcus: to remember it, we can compare it to what happens in the vital cycle of the human being. One is born, one will grow, one reproduces, and one will die. So this supracallosum sulcus is born over the most inferior part of the medial surface of the frontal lobe curved around the corpus callosum and will die or finish at the parieto-occipital fissure. One branch, or sulcus, originates posteriorly and superiorly, the paracentral sulcus, which joins at the superior portion of the brain with the *rolandic fissure*.

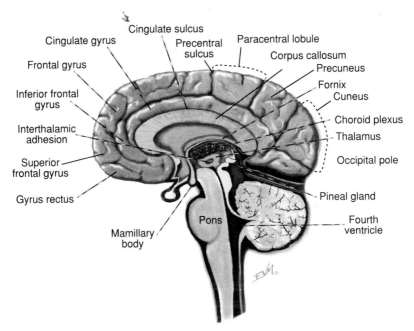

Figure 1-1 A medial view of the brain.

The second branch of the callosomarginal fissure is the *supraorbital sulcus*, just above what is known as *gyrus rectus*. Another branch, ill-delineated, is one that is born at the first curvature of the singular sulcus and extends backward and forward in many occasions reaching the upper border of the brain. Finally, it grows posteriorly and is called the *intraparietal sulcus*. Another large sulcus seen in the medial aspect of the cerebral hemisphere is the *calcarine fissure*. The calcarine fissure usually has two branches that soon join together and run horizontally forward to join the internal parieto-occipital fissure—but do not join the sulcus of the hippocampus due to the presence of the gyrus fornicatus. As the calcarine fissure runs forward, it becomes perpendicular to the internal parieto-occipital fissure. In this fashion, the medial aspect of the brain has several gyri or convolutions, including (1) the *gyrus rectus*; (2) the *cingulate gyrus*, which begins at the most anterior portion of the corpus callosum in front of the lamina terminalis, and curves around the genu and body of the corpus callosum back toward the parieto-occipital fissure. The cingulate gyrus joins the hippocampus to form with the uncus of the temporal lobe (the main part of circuit of Papez). The cingulate gyrus joins the hippocampus through the *gyrus fornicatus*. This forms the limbic system. This is the main part of the circuit of Papez. There is a limbic system for each side of the brain (right and left). The only common structure to the circuit of Papez is the *fornix*.

To continue the description of the medial aspect, in the frontal lobe, we find the superior and inferior medial frontal gyri. These are separated by the second branch of the *collasomarginal sulcus*.

Limited by the precentral fissure anteriorly and by the singular fissure inferiorly, is the *paracentral lobe*. The paracentral lobe has a motor portion *anteriorly*—which is the medial continuation of the motor strip. It also has a *sensory* portion *posteriorly*—which is the medial continuation of the postcentral gyrus, or sensory strip. Behind the paracentral lobe is the *precuneus*, which is separated from the occipital lobe posteriorly by the prominent parieto-occipital sulcus. The precuneus is also called the *rectangular lobe*. The inferior outline of the precuneal gyrus is the posterior extension of the singular sulcus.

Outlined by the internal parieto-occipital fissure anteriorly (by the superior border of the brain all the way to the occipital pole) and inferiorly by the calcarine fissure, we have in the medial aspect of the occipital lobe the cuneal gyrus, or simply the *cuneus*.

THE VENTRAL OR INFERIOR SURFACE OF THE CEREBRAL HEMISPHERE

In the ventral surface we can see the orbital aspect of the frontal lobe, the olfactory bulb, the olfactory stalk (the lateral and medial olfactory striae), and the *anterior perforated space* (see Fig. 1-2). The olfactory bulb is located in a groove known as the olfactory sulcus, which outlines laterally the inferior aspect of the gyrus rectus. Posteriorly, the olfactory stalk is divided in two, the medial and lateral olfactory striae, which form the lateral portion, a triangular zone known as the olfactory trigone. This area contains, or perhaps should be named solely, the anterior perforated space because one can see many *holes*, through which many arteries in the brain enter to supply the thalamus and corpus striatum.

The orbital aspect of the frontal lobe contains several sulci. These sulci sometimes have the shape of an X or H (see Figure 1-3). The different gyri are known as the *orbital gyri*. Medial to the large arm of the H is the *gyrus rectus*. Lateral to the most lateral long arm of the H, we find the olfactory desert. Anterior to the horizontal branch of the H is the *anterior orbital convolution.* Posterior to this horizontal branch is the postorbital convolution. The middle frontal orbital convolution continues at the frontal pole, with the most anterior part of the superior frontal convolution. It is known clinically that stimulation of the most posterior portion of the orbital aspect of the frontal lobe can create a neurodystrophic phenomenon manifested by picture-like ulcerative colitis, tachycardia, bradycardia, or irregular heart rhythm.

The mesial (inferior or bottom) aspect of the temporal lobe is the remaining ventral portion of the brain. The temporal pole is in direct contact with the most posterior portion of the orbital gyri. In that space, a portion of one *middle* cerebral artery, known as the M1 portion of the middle cerebral artery, is found. We also have two sulci, one medial and one lateral. In this fashion, the mesial aspect of the temporal lobe is divided into three gyri (medial, intermedial, and lateral).

Medially, we find the sulcus of the hippocampus. Lateral to the sulcus of the hippocampus is the hippocampal gyrus. As previously mentioned, the hippocampus forms part of the circuit of Papez. The hippocampal gyrus lies medially to the collateral fissure. The hippocampal gyrus has a hook-like area known as the *uncus*. Medial to the hippocampus, one can see the optic tract and one portion known as the *fascia dentate* situated above the hippocampal gyrus, which is separated from the hippocampus itself by the dentate fissure. In fact, as mentioned before, the gyrus fornicatus separates the calcarine fissure from the sulcus of the hippocampus. The medial striae of the olfactory stalk climbs above the corpus callosum, changes its name, and is then called the tract of Lancisi, which over the entire body of the corpus callosum, joins the one on the opposite side to form the indusseum griseum. The tract of Lancisi curves around the splenium of the corpus callosum and again changes name, and is called the fasciola cinerea, which extends down, running like a slope, forming the previously described dentate convolution (on the superior border of the hippocampus). As this fascia dentate runs anteriorly, it again changes its name and forms a small belt-like structure that curves upward over the uncus of the hippocampus, forming the band of Giacomini where it is lost. The lateral striae of the olfactory tract also extend to the hippocampus, forming a great part of the intralimbic system.

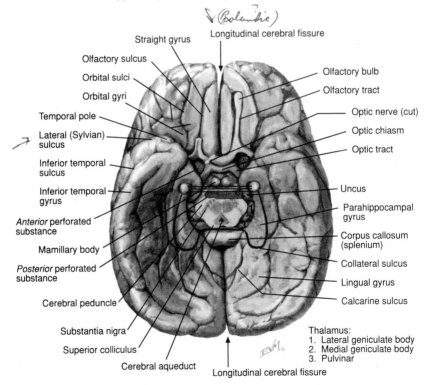

Figure 1-2 The cerebrum. An inferior view with a sectioned brain stem.

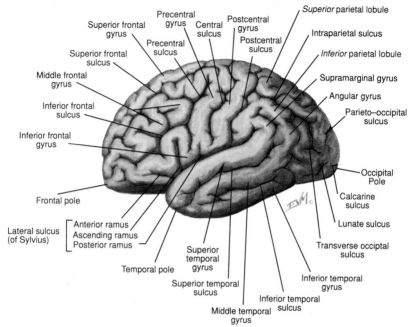

Figure 1-3 A lateral view of the cerebrum.

Returning to the description of the ventral surface of the temporal lobe, it can now be said that *lateral* to the hippocampus, we have the *collateral fissure*. We have, also, the inferior lateral fissure. As mentioned before, we have three gyri:

1. The hippocampus.
2. The medial gyri.
3. The inferior temporal gyrus.

The inferior temporal gyrus forms the inferior and lateral boundary of the cerebral hemisphere and really represents the inferior aspect of the third temporal convolution, as seen in the lateral aspect of the brain.

Between the inferior temporal gyrus and the hippocampus, one can find the *fusiform gyrus*, which is limited externally by the inferior temporal fissure and medially by the collateral fissure. The most posterior portion of the fusiform gyrus has the shape of a tongue, and it is called the *lingual gyrus*, which extends backward forming integral part of the *occipital* lobe.

A lesion at the most posterior part of the lingual gyrus produces a peculiar syndrome consisting of the patient abruptly feeling that he or she has lost his or her hand and cannot find it. This is known as the syndrome of the lost hand of Oppenheim.

In the ventral aspect of the brain, one can also see the pituitary stalk, the optic chiasm, the tuber cinerium, mammillary bodies, the posterior perforated space, and the cerebral peduncles, which will be described subsequently.

The Frontal Lobe 2

As mentioned in Chapter 1, the external surface of the frontal lobe, as well as the parietal, occipital, and temporal lobes have two sulci and therefore, three gyri. Consequently, it will be easy to remember that with two sulci there are three convolutions, superior, medial, and inferior. This will allow those who review the anatomy to remember it; this is the easiest way to remember the lobes (see Figure 2-1).

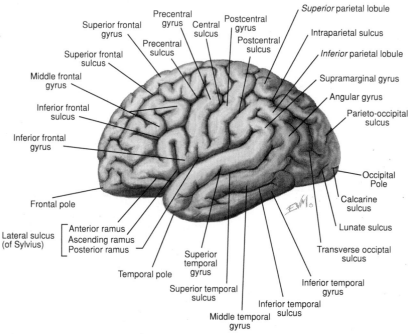

Figure 2-1 A lateral view of the cerebrum.

The frontal lobe, at its external surface, has a *superior* frontal fissure and *inferior* frontal fissure. This divides, therefore, the frontal lobe and three convolutions, or gyri: superior frontal gyrus, middle frontal gyrus, and inferior frontal gyrus. Each one of the sulci, the superior frontal fissure as well as the inferior frontal fissure, subdivide in an ascendant and descendent prolongation creating what is called the *precentral sulcus* (some anatomy books state that the frontal lobe indeed, has three sulci and therefore, four convolutions: superior, medial, and inferior convolution, and the fourth one, the motor strip, which is located in front of the central sulcus, or *fissure of Rolando*). The precentral fissure is frequently interrupted by some gyri. From the precentral sulcus, the superior, the middle, and frontal convolutions extend forward and downward toward the pole of the frontal lobe to curve ventrally toward the base of the brain, forming the so-called orbital portion of the frontal lobe. The motor strip, also called the *ascending frontal convolution*, is located in between the prerolandic and Rolandic fissure, and extends from the sylvian fissure to the superior border of the brain. The inferior frontal convolution is located between the sylvian fissure below and the inferior frontal sulcus above. It contains the *Broca's area*, which is the center of the articulate speech. Within the frontal lobe, there is a wide range of areas. They have to be considered from the morphologic as well as from the functional point of view. Here in the frontal lobe lies the *primary motor cortex*, which is known as areas **4** and **8**. There are also supplementary motor areas that are in area **6**, which is primarily the inferior portion of the frontal lobe. If a person has damage of the motor strip where the pyramidal tract is originated, he or she will have contralateral hemiparesis or hemiplegia. According with the anatomy as well as the cortical representation of the different part of the body (see *humunculus*), the medial part of the frontal lobe—which is the motor portion of the paracentral lobe—controls the movements of the toes, ankles, and knees. At the convexity, in the motor strip, is the control of the motor ability of the hip, trunk, shoulder, elbow, and wrist (see *humunculus*). At the junction of the upper-third with the mid-third, and including the mid-third, the control of the hand, thumb, and ring finger is located.

At the lower third of motor strip and the junction with the most posterior portion of the inferior frontal convolution, we have the *Broca's area*, which represents articulate speech. This is the area of vocalization. This area **4**, or motor strip, is located in front of the central sulcus and occupies the greater portion of the precentral gyrus. Area **4** is characterized by a six-layer pattern that is typical for the neocortex.

The neocortex layers are:

- The stratus molecularis
- The stratus granularis externa
- The stratus pyramidalis; which contains the large Betz cells
- The stratus granularis interna; also populated by large amount of granular cells
- The stratus ganglionaris; which has large and medium size Betz cells
- The stratus multiformis

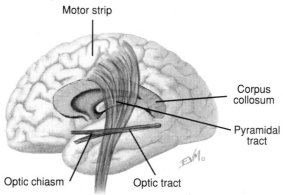

Figure 2-2 Two pyramidal tracts.

We resume the description of the pyramidal tract as originating from the motor cortex, extending into the white matter backwards, and descending down to the posterior limb of the internal capsule where the cortical bulbar fiber occupies the genu of the internal capsule (see Figure 2-2). It can also be said that the area of the trunk also extends from the genu of the internal capsule to the most anterior part of the posterior limb of the internal capsule.

The upper extremity is located in front of the area that controls the motor ability of the trunk at the posterior limb of the internal capsule, and finally, the legs are in the most posterior portion of the posterior limb of the internal capsule. Therefore at the internal capsule, the fiber that brings the control and motor ability of the foot are anterior, then the upper extremity, then the trunk, then the lower extremities from cephalad to caudal direction. The same organization is followed in the midbrain; therefore, the foot is more medial (in front of the substantia nigra), which will be explained subsequently. Then follows the upper extremity, the trunk, and lower extremity from a medial to a lateral direction. The same organization is followed at the pons and also at the medulla before the pyramidal tract decussates (see Figure 2-3).

The main pyramidal tract, or the main bundle of the tract, is joined by the cortical bulbar fiber from the frontal eye fields located in front of the motor strip (see *humunculus*). This fiber constitutes the most medial and dorsomedial portion of the pyramidal system at the cerebral peduncle (notice the *leg*, *arm*, and *face* distribution in the fibers). At the lower border of the lower third of the medulla are 70 percent to 85 percent of the cortical spinal fibers, which usually decussate and continue caudally in the spinal cord, which constitute the lateral cortical spinal tract. In this decussation, the upper extremity fiber crosses more rostrally and the lower extremity more caudally, and relates to the lower extremity movements. Those fibers concerned with the upper extremity movement are medial in the tract. The majority of the pyramidal fibers that do not decussate at the motor decussation extend into the spinal cord and the tract is known as the *ventral cortical spinal tract* (see Fig. 2-3). Most of these fibers decussate in the anterior white commissure before terminating. This fiber carries, most particularly, motor impulses for voluntary control of the neck and trunk muscles.

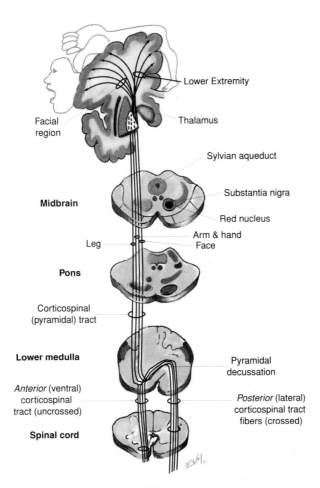

Figure 2-3 The *corticospinal pyramidal tract.*

This is a very important factor that will subsequently be discussed when we review adversity seizures. The sternocleidomastoid muscle has an ipsilateral nerve supply from the cerebral cortex. The destruction of the motor cortex or the corticospinal pathway will result in a loss of fine voluntary movements, especially of the hand and foot. This loss tends to be a permanent motor defect when there is lesion in this system. Although if the patient develops hemiplegia, he or she might learn how to walk, but the performance of fine motor activities of the extremities will be very difficult.

There are secondary motor areas at the base of the central fissure, which coexist with the second sensory area at the base of the precentral and postcentral gyri. Area **6** lies rostrally to area **4** and is also in the medial portion of the hemisphere at the frontal lobe. It is separated from the area **4** by a very small area known as the area **4S**. Area **6** is subdivided into two portions, one that lies parallel to the motor cortex and another, more rostrally, that occupies the superior frontal gyrus in both the lateral and the medial aspect of the brain. This controls the contralateral lower extremity, trunk, upper extremity, as well as the face.

Area **6** has a connection to the adjacent cortical areas and also through the cingulate gyrus.

There are also connections to the putamen and globus pallidus to the frontal lobe—connections that are responsible for grasping, sucking, rooting, and snouting. Therefore, a *frontal lobe tumor* or lesion could produce grasping reflexes when someone touches the palm of the hand. The primitive motor functions are under the control of the basal ganglia, which also control the sucking reflex.

As the myelinization of the pyramidal tract occurs, these functions become dormant, but if there is a destruction of this portion of the cortex that connects with the basal ganglia, the inhibitory activities that take place over the basal ganglia disappear. Therefore, a patient with a frontal lobe lesion could present with involuntary hand grasping as well as a sucking reflex due to the lack of cortical inhibitory effect.

For the purposes of memorization and understanding, let's say that the baby moves in the uterus because of the basal ganglia (which are responsible for the grasping and sucking reflexes). When he or she is born, he or she will suck or grasp (a function of basal ganglia) even though there is development and some myelinization of a pyramidal tract, it is currently not myelinated. When the *pyramidal tract* becomes myelinated, it is the "boss" and will look down to the *basal ganglia* and say in authoritative voice, "Friend, I don't need to suck, I can chew; I don't need to grab, I can walk." However, if the motor strip becomes ill, the basal ganglia will look up to it and say, "What is your fuss, you say that you are the boss, but you are not, I can suck again and grasp again." That is why a frontal lobe tumor returns a patient's primitive functions.

A patient who has a lesion in area **6** could present with *adversative seizure*, manifested by jerking movements of the face, arm, and leg on the **contra**lateral side of the lesion—with the head deviated also to the *right*. Here, one should remember that the sternocleidomastoid muscle is supplied by the ipsilateral hemisphere. If the lesion is in the *left* frontal area **6**, the jerking movement will occur in the *right* side of the face, arm, and leg, and the movements of the head will go to the right (the same side of the seizure). This is because the muscle that is contracting is the *left sternocleidomastoid*, and the head will move to the *right*. There is the erroneous general conception that every muscle in the right side of the body is supplied by the left hemisphere. If this was the case, which it is **not**, the head would move toward the left when the right sternocleidomastoid contracted. The truth is that the left hemisphere supplies the left sternocleidomastoid, therefore the head moves to the right, contrary to the direction one expects if the right sternocleidomastoid muscle is supplied from the left, hence the name of ***advers**ative seizures*.

Due to the connection between the putamen and the globus pallidus with the cortex through the corticostriatum pathway, some of those patients with frontal lobe lesions present with abnormal movements that simulate a contralateral hemi-Parkinson's. A meningioma of the convexity of the frontal lobe—this lobe also simulates a contralateral Parkinson's. The prefrontal eye fields number 1 and 2, which are located in front of the premotor cortex, have a direct connection with the nucleus of the VI cranial nerve and paraabducent nucleus. This explains why the eye moves contralaterally to the side of the lesion.

It could be said that stimulation of the frontal eye field, which is in front of the premotor cortex, produces *conjugate horizontal deviation* of the eyes to the opposite side. In cases where there is a lesion in this area, there is a forced deviation of the eyes opposite to the lesion. This phenomenon could also occur in lesions of the inferior frontal gyrus. Lesions in these areas might, at times, simulate nystagmus. However, these eye movements are pendular in nature, and many times, it is difficult to state for certain using the naked eye, if we are dealing with a *true* nystagmus. If the eyes are tonically deviated to the opposite side of the lesion, one could obtain doll's eye movements and caloric responses, because they are preserved.

"COWS"—Cold temperature in an ear will cause movement of the eyes to the opposite side, and warm temperature to the same side.

The pars opercularis and pars orbitalis of the inferior frontal convolution are bordering a big lobe known as the *island (insula) of Reil*. As mentioned before, stimulation of the orbital aspect of the frontal lobe can produce a neurodystrophic phenomenon characterized by tachycardia or bradycardia, picture-like ulcerative colitis, and hyperventilation. The orbital surface of the frontal lobe receives the inferior thalamic radiations coming from the anterior and dorsomedial nucleus of the thalamus (see Figure 2-4). The orbital portion of the frontal lobe has wide communication with the temporal lobe at the amygdaloid nucleus through the uncinate fasciculus (the ventral division of the uncinate fasciculus).

Many of the fibers of the uncinate fasciculus not only connect the amygdala with the frontal lobe, but also with the hippocampal gyrus. Other fibers interconnect the superior temporal gyrus with the frontal cortex, particularly with the inferior and medial frontal gyri. This portion of the uncinate fasciculus is known as the dorsal division of the fasciculus and occupies the anterior limb of the internal capsule. As it will be discussed, the uncinate fasciculus on its dorsal portion joins the fronto-occipital fasciculus at the capsula extrema.

This association bundle that exists between the frontal and temporal lobes may, on occasions, produce ortho- and antidromic impulses which may give, to the examiner, the wrong interpretation in the location. For example, a frontal lobe tumor, by transmitting the irritation through the uncinate bundle, could create visual hallucinations that make the examiner suspicious of a lesion in the rostral part of the temporal lobe. Contrary-wise, a temporal lesion might simulate a frontal lobe syndrome.

In the medial aspect of the brain, at the most upper and posterior portion of the superior and medial frontal gyrus, cases of *akinetic mutism* have been

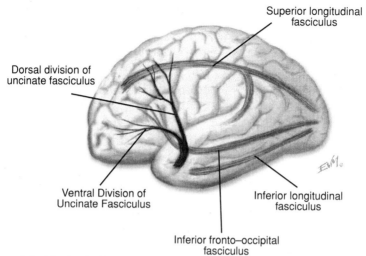

Figure 2-4 The fasciculi.

described. Cushing described the first case of akinetic mutism in a patient who had a paracital meningioma compressing the most posterior portion of the superior and medial frontal gyrus.

The frontal lobe has a wide connection with the contralateral cerebellar hemisphere via the frontopontine tract, and extends from the premotor cortex through the internal capsule to the ipsilateral nucleus of the pons, and from there, through the middle cerebellar peduncle to the cerebellar cortex. The cerebellar cortex has an efferent pathway to the cerebello-dento-rubro-thalamo-cortical tract. Therefore, lesions that stimulate the frontopontine tract will, of course, produce stimulation of the cerebellum. Overstimulation could produce cerebellar fatigue and consequently, *cerebellar dysfunction*. Hence a frontal tumor can cause a presentation of ataxia. When this ataxia originates in the frontal lobe, it is known as *Brun's ataxia*.

In cases of frontal lobe tumors or lesions, the patient could present, therefore, with right hemiplegia or hemiparesis, Broca's aphasia or dysphasia, and/or contralateral tremors that mimic hemi-Parkinson's. If the irritation occurs in the area of the frontal lobe that controls eye movements, a nystagmus-like picture could appear. Adversity seizure could also be a presentation. Neurodystrophic phenomenon has been seen if the lesion is in the orbital aspect of the frontal lobe and a clinical presentation might be one of a picture-like ulcerative colitis and/or impairment of the heart rhythm. These patients have passive behavior and a flat affect, or they may abandon their personal hygiene. Some of these patients could be confused as depressed individuals. At times, this frontal lobe syndrome is accompanied by inappropriate laughing (gelastic seizures due to the wide connection that exists in between the orbital aspect of the frontal lobe and the hypothalamus via the middle forebrain bundle. In meningiomas of the olfactory groove, in addition to the other mentioned symptoms and signs, the patient could present with atrophy of the ipsilateral optic nerve. Due to the increased intracranial pressure created by the tumor itself, however, that same individual will have contralateral papilledema (Foster-Kennedy syndrome). The lectors should remember that this syndrome is not always produced by a tumor. There is a case report (reported by another author, Dr. Jose Biller, and this author) of Foster-Kennedy syndrome produced by perichiasmatic arachnoiditis.

The Parietal Lobe 3

The parietal lobe contains the somesthetic cortex known as areas **1**, **2**, and **3**; the sensory association cortex, areas **5** and **7**; and sensory speech and sensory cortex. On its external surface, the parietal lobe presents the previously described sulci. It should be remembered, by a previous description, that there are two sulci and three convolutions.

To refresh the previous description, let's repeat that the *intraparietal sulcus*, which begins in the most posterior portion of the sylvian fissure, is found just behind the central sulcus. It extends upward in a parallel course to the rolandic fissure, and at the mid-third of the parietal lobe, it turns backward extending to the external parieto-occipital fissure. Where the intraparietal sulcus makes its bend, the ascending portion of the sulcus can be seen. This ascending portion, together with the first-third of the parieto-occipital fissure, creates the postcentral sulcus or postrolandic fissure. The parietal lobe, therefore, is divided into three convolutions. The postcentral gyrus, or ascending parietal convolution, is bound in front by the rolandic fissure and posteriorly by the postcentral sulcus.

Limited anteriorly by the postcentral sulcus, posteriorly by the external parieto-occipital fissure, and inferiorly by the sylvian fissure, is the superior parietal lobe. This superior parietal has some of its portion extending towards the occipital lobe. The inferior parietal gyrus is located between the intraparietal sulcus superiorly, sylvian fissure inferiorly, and external parieto-occipital fissure posteriorly. It is subdivided in two portions by a very ill-defined sulcus, the so-called *gyrus supramarginalis*, just immediately behind the first portion of the intraparietal sulcus. Some authors call this the *sigmoid* gyrus. Immediately behind it is the *angular gyrus* (which joins with the *supramarginal gyrus* anteriorly). The angular gyrus, posteriorly and inferiorly, has a direct continuity with the posterior portion of the middle temporal convolution.

The medial surface of the parietal lobe has a convolution with a rather small, rectangular shape, known as the *quadrate lobe*, or *precuneus*.

The somesthetic cortex of the postcentral cortex has six layers:
1. Molecular layer
2. External granular layer
3. Pyramidal layer or stratum pyramidalis
4. Internal granular layer
5. Ganglional layer
6. Layer of lamina multiformis

Along the posterior wall of the rolandic fissure, at the depths of that wall, the cortex has loosely arranged small cells known as *koniocortex*. These pick up stimuli generated when the skin is stimulated. The cortical arrangement of the somesthetic cortex has medially at the paracentral lobe, the representation of the *tactile* sensation for the scrotum, foot, leg, and knee. It is represented at the most superior and medial portion of the sensory strip, from medial to lateral, and localizes portions of the leg, hip, trunk, neck, and shoulder (similar to the motor representation of the humunculus, but this is sensory). In the descending portion of that cortex, at the mid-portion, the forearm, hand, and fingers are represented. Immediately after the fingers, at the junction with the inferior third of the somesthetic cortex, we find the sensory representation for the thumb, eyelid, face, and upper and lower lips. At the bottom of the somesthetic cortex, we find the localization areas for the tongue and pharynx. It should be mentioned that the representation of the *fingers* is larger than the one for the thumb.

If one destroys or ablates the somesthetic cortex, the person loses sensory, epicritical (distinct touch) sensory recognition. The sense of position is lost as well as refined temperature gradation recognition. The vibratory sense also is impaired. This portion of the cortex is the one that receives the outflow of the ventro-posterolateral (VPL) nucleus of the thalamus. Clinically, the vibratory and position senses are more deeply impaired in a lesion of the postcentral gyrus than in a lesion of the VPL nucleus of the thalamus. It could be said that the thalamus would always hear the tuning fork.

The Betz cells (motor cells) found at the *sensory* cortex contribute immensely to the muscle pump for the venous return of the lower extremities. Lesions at the most anterior portion of the sensory strip tend to cause venous stagnation and hence, *thromboembolism*. A patient with a *parietal lobe lesion*, for this reason, tends to have a *pulmonary embolism*. A physician should be alert to this condition when he or she operates on the parietal lobe.

The entire parietal lobe is related to *proprioceptive* stimuli. Concerning the gustatory area, the parietal lobe cortex perceives them at the most inferior portion of the postcentral gyrus as it joins the insula of Reil.

The somesthetic cortex has:

1. Association fibers that connect anteriorly to the motor cortex just below the rolandic sulcus.
2. There are also connections with the visual cortex.
3. Fibers to the temporal lobe at the most inferior portion.
4. There are also efferent fibers that extend through the internal capsule to the red nucleus.

Finally, these association fibers interrelate with the contralateral parietal lobe through the corpus callosum.

The sensory radiation includes both fibers from the somesthetic cortex to the thalamus, and from the thalamus to the somesthetic cortex. The connections of the cortex are primarily extending to the dorsal thalamus and to the VPL (ventro-posterolateral nucleus) and VPM (ventro-posteromedialis nucleus).

The fibers that come from the thalamus, *efferent thalamic fibers*, arise from the ventro-posterolateral nucleus through a very large bundle of ascending sensory fibers. Also, fibers from the VPM carry *proprioception* as well as *tactile*, *pain*, and *temperature* sensations. They are located at the posterior limb of the internal capsule. Stereoception is, therefore, represented at the sensory strip. The somesthetic cortex also has connection with the basal ganglia at its most external portion of the lenticular nucleus.

Tumors or lesions in the parietal area at the sensory cortex, produce *astereognosis*. The patient is incapable of recognizing any objects placed in the hand opposite to the lesion. The patient, however, is capable of feeling the object but cannot recognize it. They could also have *agraphesthesia*, or inability to recognize with the eyes closed, letters written on the skin of the contralateral side of the body.

An interesting phenomenon is when a patient who has a parietal lobe lesion, he or she might feel a touch in the extremity *opposite* to the lesion, as well as in the extremity that is not affected (normal side). If the stimulus is *simultaneous*, meaning touching at the *same* time, for example both arms, the patient who was able to feel the stimuli previously tends to ignore or neglect the tactile sensation opposite to the lesion. This is called *extinction*.

A patient affected by a lesion located at the *parietal association area* may have a very interesting phenomenon known as *apraxia*. The following is a description of the phenomenon of apraxia. When a patient feels an object in the palm of the hand, the brain might not recognize it. The patient has a phenomenon similar to astereognosis; we can call it agnostic apraxia. For example, the patient wishes to comb his or her hair. In order to do this, the brain, at the same time, will have to "think" of all the skillful voluntary movements to do the combing. Obviously, he or she also has to have the ability to recognize the object. If he or she now wishes to make the actual movements, he or she needs the motor strip. Such an area is located in the *motor association field* in front of the motor cortex, at the mid-third of the middle frontal gyrus. This is called the *kinetic* center. There is a connection between the center of the somesthetic cortex (that feels the object) and the kinetic center. The patient's mind establishes a formula of how he or she is going to comb himself or herself (with a hairline, natural style, etc., etc.). This is the *Lipman formula*, or *idiokinetic* formula. If there is a lesion in between these two centers, the patient is incapable of imitating. If one tells the patient, "make the sign of the cross like me," or "comb your hair the same way as your neighbor patient," he or she will have difficulty. If the patient is told "touch your nose like me," with implication that he or she should imitate the examiner, he or she will not be able to do it and will be clumsy when performing that movement. As a result, one might equivocally state that the patient has an *ataxia*—when indeed what he or she has is an *idiokinetic apraxia*. That is why if you ask a patient to imitate the examiner's writing or drawing of a daisy or a square, the patient will not be able to build on what was already drawn on a paper. This is called *constructional apraxia*. It might create a sensation of mystery for the person who hears this word for the first time but in reality, it is nothing else than an *idiokinetic* apraxia. At times, the patient will be able to do a *single* movement but is unable to perform a sequence of the same single movement (i.e., he or she will comb his or her hair once but will not be able to do it again).

Patients with lesions in the areas of association in the parietal lobe, particularly in the *superior* parietal lobe, may have misconceptions of the geography or scheme of their bodies. They might not recognize the nose, ears, mouth, shoulder, and other body parts. If such a patient is asked to "show me your nose," he or she might show you a fist, stating "yes, yes, this is my nose." This phenomenon is known as *dysautotopognosis*. He or she might deny a part of his or her body and if asked "does this arm belong to you?" He or she might say "no," or "it belongs to someone else." This phenomonen is known as *asomatognosia*. Sometimes, patients like the one just described, may take a shower and not clean or wash the affected part of the body. The patient could also add an extra half, which could be the examining physician, and even state "you slept beside me the whole night." Doctors are advised, therefore, to go with the nurse during the rounds to avoid accusations from the patient. In a patient that has an amputation of a limb, the somesthetic cortex will continue for awhile to have the representation of that limb. The patient might still feel a leg that he or she does not have; this is the *phantom limb* phenomenon.

There are association areas between the occipital lobe and the parietal lobe. At the junction of the posterior parietal with the occipital lobe, we have an *optokinetic center*. Its main function is to pursue objects when they move slowly. That is why it is also called the *slow pursuant system*. This is why a person could have a sensation of *vertigo* when looking at objects through the window of a car that is moving at a high speed. This *slow pursuant* system is not able to "handle" objects that are moving at a *high* speed. Therefore, the person develops *nystagmus*. The *inferior parietal lobe*, as mentioned before, has two portions—the *supramaginal* and the *angular gyrus*. The supramarginal gyrus is surrounding the most posterior portion of the sylvian fissure and its main function is primarily an association function.

The supramarginal gyrus is also associated with *visual and auditory* impulses from the occipital and temporal lobe. Immediately below that cortex, there is a bundle known as the *inferior occipitofrontal longitudinal fasciculus*, which contains fibers that interconnect the occipital and parietal lobe with the frontal lobe (see Figure 3-1). In the same bundle are fibers that interconnect the *auditory* areas with the auditory area of the opposite side.

The *angular gyrus* is an extremely important part of the human brain. With a lesion in the dominant hemisphere, which allows the understanding and comprehension of the written and spoken language, the patient might not be able to read. This condition is known as *alexia*. The patient will present with an inability to write, *agraphia*, or to even read whatever he or she might have written. He or she will have an inability to understand spoken language (*sensory aphasia*). *Acalculia* is a defect that presents in lesions at the angular

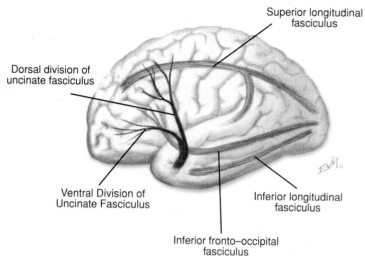

Figure 3-1 The fasciculi.

gyrus. This patient will not be able to subtract, add, or multiply, in other words, he or she will not be able to balance a checking account. However, he or she can talk, but answers or conversation might be absolutely unrelated to the spoken subject. This is also known as *"fluent aphasia."*

In summary, a patient who has right-to-left disorientation, finger agnosia, impaired touch, vibration or position sense, astereognosis, agraphia, alexia, asomatognosis, dysautotopognosis, acalculia, possible idiokinetic apraxia has a parietal lobe syndrome known as *Gerstman syndrome*. The patient also could have sensory aphasia. He or she may even be able to read but cannot understand what was read (*Wernicke's aphasia*). Sometimes, as mentioned before, the patient can have edema in one extremity—opposite to the lesion.

Language, the Mechanism of Speech, and Aphasias 4

The language is the means by which the human beings are capable of expressing their thoughts through the oral language (expressive speech) or by written means. When a person speaks, he or she expresses his or her thoughts. When he or she hears or reads, he or she understands the thoughts of the other. Speech begins to develop at a very early age and it is, therefore, very advisable to speak to an infant even though it is thought that the infant does not understand. Thousands of neurons are "wiring themselves," establishing connection that will, in the future, represent the mountain of knowledge that we learn through life.

The *mechanism of the speech* is a rather complex phenomenon. In order to express our thoughts, several steps take place within the brain. First of all, a person has to have a concept or idea (the intellectual part of the speech). Second, a person must evoke from memory the word that corresponds to the concept that he or she wants to express. And third, the person has to pronounce, or articulate (motor portion of the speech), the word.

On the other hand, when a person listens, to understand what is spoken, he or she must first listen to the word; second, he or she must identify the idea that was directed to him or her; and third, must comprehend the meaning (intellectual part of the speech).

Certain *cortical lesions* can create difficulties or impossibilities for the patient to understand or express himself or herself, whether it is written or orally. This disturbance constitutes the *aphasias*. Aphasia, therefore, is a disturbance of the speech that will have several specific characteristics.

The aphasic person is one that without suffering paralysis, blindness, or deafness, is *incapable of expression* by written or spoken language and/or *comprehension* of the words or the written subject. Since we have already discussed the frontal and parietal lobe, this is the most appropriate time to explain aphasia. In the cerebral hemisphere, there are four specific primary areas that correspond to the four varieties of language.

1. Motor speech—expressive oral and written.
2. Center of comprehension—oral and written language.

This implies that there is a specific motor center for spoken and written language. A motor center that allows us to think and utilize the musculature to do what is needed; and a center that "thinks" and includes the components necessary to write, in other words, to use the muscles of the hand. Therefore, we could say that the center of spoken language is directly connected to the lips, tongue, larynx, and the hand.

Sensory speech also has two centers.
1. The one that corresponds to the memory of *spoken* language.
2. The center of the memory of *written* language.

As mentioned previously, the motor center that controls the movements of the lips and hands joins with the *sensorial* center of speech, and then subsequently connects with the center of hearing and vision. This motor aphasia is known as *Broca's aphasia*. In this situation, the patient is not able to speak. However, an interesting phenomenon is that many of these patients can sing along with you certain songs that were learned in the remote past. If a physician wishes to explore the inability in this type of a patient, one can show him or her, for example, a pen and ask "what is this?" He or she may not be able to speak, however, by using a series of questions, and asking him or her to answer with an affirmative gesture of the head, he or she can indicate when the word that the explorer used corresponds to the object that is shown. The explorer should proceed with questions like "is this a comb?" or "is this a gun?" Finally, the patient is asked "is that a pen?" and he or she will probably answer with an affirmative gesture. At times when the patient answers, if he or she is not totally aphasic but dysphasic, he or she may *substitute* the name of the object with a different word (*paraphasic error*), or he or she may make a deformation of the word (e.g., a pencil is called the "pencomb"). If the object shown is a watch, he or she may call it "catch" instead of watch.

At times when the patient speaks, there is a disturbance of semantics or grammar. He or she speaks in a jargon language. An examiner can explore the patient's inability by asking him or her to repeat simple words many times. The next sentence should contain more complex words, and then, simple sentences. The fluidity of the speech and the paraphasic errors, or substitution of words, can also be observed.

In the exploration of the level sensory speech or comprehension of spoken language, we can begin by asking the patient to execute simple movements like "open your mouth," or "stick out your tongue." He or she can be asked to "put your right hand over your left shoulder." Then, the examiner can observe how well the order is executed. Sometimes, it is very obvious that the patient does not have an understanding of the question. The examiner should be certain that none of the commands carry any danger for the patient when executed. The patient with *sensory aphasia* loses the concept of fear. I have seen an aphasic patient who was ordered by a friend to enter a corral with ferocious bulls, and he went in. It "never occurred to him that the bull could kill him." The other aspect of the sensory aphasia is the inability to understand written language. This is called *alexia*. It can also occur when the patient is asked to read written music. He or she has *amucia*. This is indeed what one might call "blindness in comprehension." This type of aphasia is known as *Wernicke's aphasia*. This patient, contrary to what happened in the person affected with Broca's aphasia, will be able to speak with a rather irrational language but will speak a lot (*logorrhea*). For example he or she may say, "Yeah the car, the football, . . . it is the dinner, the shoe . . ." It is a lot of jargon with no meaning, and it proceeds incessantly (logorrhea).

At times the patient could have a mixed aphasia, motor as well as sensorial, or in other words, he or she has *anarthria* (cannot articulate) plus Wernicke's aphasia. This patient can neither comprehend nor speak, however can copy a given sentence. The etiology of this aphasia can be due to vascular occlusion disease, trauma, tumors, and/or infection like abscesses.

The Occipital Lobe 5

The occipital lobe occupies the most posterior portion of the cerebral hemisphere. It has three aspects: medial, inferior, and posterior. It contains two very important areas: the visual cortex (area 17) and the visual association cortex (areas 18 and 19).

The occipital lobe is divided in its external surface by two distinct sulci in three convolutions or gyri: the superior, middle, and inferior. The middle occipital gyrus is connected with the angular gyrus as well as to the middle temporal convolution by the supernumerary gyrus.

The *inferior occipital convolution* has a direct connection to the inferior temporal convolution by another supernumerary gyrus. The mesial portion of the occipital lobe presents the calcarine fissure. There is a triangular gyrus known as cuneus that is situated between the internal parieto-occipital fissure and the calcarine fissure and extends anteriorly at the most posterior portion of the corpus callosum. Area **17** receives the axons from the ganglion cells of the retina, along the optic nerve and optic tract. The optic tract stops at the lateral geniculate body and from there, runs to area **17**. All of these neurons from the lateral geniculate body transmit impulses along the geniculocalcarine pathway or *optic radiation* to the upper and lower lips of the calcarine fissure. **Area 17** has the primary function for *vision*. Due to the decussation of the optic pathway at the chiasm, which will be explained later on, each area 17 receives stimuli from the contralateral half of the visual field. If it is destroyed unilaterally, area 17 will create a visual field defect known as *complete homonymous hemianopsia*. **Areas 18** and **19** are secondary and tertiary association areas and their main function is the visual recognition or visual recall. We also have **area 39**, which is an association area for the symbolic languages. This area is not responsible for reading, rather for the comprehension of what is read. The angular gyrus is intact, therefore the reader does not have a true alexia, but a visual agnosia for the symbols (or letters).

The destruction of the area 17 is usually due to an embolic phenomenon in the *posterior* cerebral artery. If both areas 17 are destroyed, one sees a very peculiar deficit known as *Anton's syndrome*. Even though the patient is totally blind, he or she is not capable of recognizing that he or she is blind. Usually, it is due to destruction in between the thalamus and the visual cortex (area 17).

On the other hand, in a lesion that occurs in area 18 of the dominant hemisphere, one can encounter no blindness, but rather an inability to recognize objects and symbols like letters, numbers, and so on. The lower lip of area 18 is primarily concerned with symbolic inanimate objects. The superior lip is concerned with animate objects, especially with the body itself.

Area 19 is primarily concerned with the revisualization of objects. This area adjoins the angular gyrus. Lesions in areas 18 and 19 produce the visual agnosis. For example, a patient who sustained a gunshot to the head, which did not penetrate the skull but contused the occipital lobe at area 19, does NOT recognize people. I have had a patient who sustained such a trauma. This patient was lying in bed and did not seem to mind if anybody entered the room. His wife called me telling me, "my husband does not love me anymore, he is not eating, urinates in bed, and does not leave the room." I instructed her to go with me to the room without saying a word. It appeared as "Mrs. Nothing" entered the room. She was about to burst into tears when I signaled her to talk to him. She said "Hi John." The patient immediately recognized that it was his wife talking to him. It was the voice that allowed him to know that it was his wife, although he could not recognize her when he was looking at her. The patient broke into a very lengthy conversation, asking "Mary, where have you been, I was anxious to see you?" This continued on and on and on. The wife asked "why don't you eat?" To which he responded "what food?" There was a food tray right in front of him. She told him "take the glass" and he could not recognize it, until she put it in his hand, and he said "oh, oh, it is glass." Yet, he did not know if the glass contained milk or orange juice by looking at it until his wife put it in his mouth, and then stated "oh, it is milk, it is milk, give me more, give me more." It appeared that the patient was now "on" and he could continue to speak with the wife. The wife asked him "why don't you leave the room to walk outside?" It was obvious that he did not know what the door was for. When she asked a new question "why don't you go to the bathroom instead of urinating in bed," he stated "we have no bathroom." He was obviously ashamed, at that moment, that he had been urinating in bed. This was a very sad example of what visual agnosis for animate symbolic and nonsymbolic objects can produce in a human.

If the connection in between the *thalamus* and areas **18** are destroyed, a syndrome in which the patient does not recognize letters or numbers can be created. Therefore, he or she has a syndrome of *alexia without aphasia* or perhaps, it is better to state that the patient has a visual agnosia for inanimate *symbolic* objects.

Areas 18 have intimate connections with areas 17 and 19 at the same hemisphere by short association fibers, and also with the contralateral hemisphere by commissural fibers via the corpus callosum. Areas 18 receive fibers from the frontal lobe and also have auditory association areas to the tip of the temporal lobe and insula.

These fibers contribute to the superior longitudinal fasciculus and inferior fronto-occipital bundle. The superior longitudinal fasciculus contains primarily fronto-occipital fibers but also interdigitates with the fibers of the corpus callosum. The superior longitudinal fasciculus runs on top of the superior border of the insula–separated from it by fibers of the corpus callosum. The inferior occipito-frontal fasciculus connects the entire visual cortex with the orbital aspect of the frontal lobe, and laterally it connects to the lenticular nucleus and to the external capsule.

The *inferior longitudinal fasciculus* is an association bundle that interconnects the occipital cortex, the cuneus, and the lingula with the superior, the middle, and inferior temporal gyri.

The Temporal Lobe 6

The temporal lobe, it could be said, is that part of the brain that communicates the outside world to us—since it is the center for hearing, smelling, taste, and visual representation. The temporal lobe is limited superiorly and externally by the sylvian fissure. It has a lateral and inferior, or mesial, aspect. The outer surface is divided by two sulci in three convolutions. The first temporal sulcus runs parallel to the sylvian fissure and extends from the tip of the temporal lobe to the parietal lobe. The second temporal sulcus, also called *inferior sulcus*, also runs parallel to the previous one but contrary to the first one, is interrupted by several gyri. As stated before, the temporal lobe, therefore, has a superior, medial, and inferior temporal gyri or convolutions. The superior temporal convolution is located between the sylvian fissure above and the superior temporal sulcus below. It extends posteriorly into the supramarginal convolution. The second temporal gyrus is located in between the superior and the inferior temporal sulci and continues posteriorly with the angular gyrus and middle occipital gyrus. The inferior temporal gyrus is located below the inferior temporal sulcus and it extends almost in continuity, posteriorly, with the inferior occipital gyrus.

The inferior, or mesial, aspects of the temporal lobe present two sulci. The first extends from the occipital pole to the tip of the temporal lobe anteriorly, and is many times divided by a small interpost gyrus. Medially, the temporal lobe is separated from the hippocampal gyrus by the *collateral fissure*. Laterally, it extends to the lateral surface of the brain. From the inferior border of the brain to the midline the convolutions include the external occipitotemporal gyrus and the lingual lobe, located between the calcarine fissure above the most posterior part of the collateral fissure below. The lingual lobe continues anteriorly with the hippocampal convolution. Between the inferior temporal gyrus and the hippocampal gyrus is found the gyrus *fusiform*. In the second temporal convolution is the acoustic center, areas **21** and **22** of Brodmann. Audition is the most important function of the temporal lobe, consequently, it is an important part of language. The *primary auditory cortex* is located at the most medial and superior portion of the first temporal gyrus.

Audition has bilateral representation. The secondary auditory areas are located in the areas **41** and **42,** as described previously. Lesions in the most posterior portion of the temporal lobe produce difficulty in recognition of spoken language. *Unilateral* temporal lesions might result in difficulty in localizing a sound in space.

Although there is a specific center for music, a lesion in the most *anterior* portion of the first temporal gyrus can also impair the ability to recognize music (amucia). Certain musical passages can produce musicogenic seizures (a particular song may induce a seizure).

A lesion in the temporal lobe can also affect optic radiation, producing a visual field defect. The hippocampal gyrus has functional olfactory as well as gustatory perception areas. The amygdala constitutes a very important nucleus that has several subdivisions and it forms a very important part of the tip of the temporal lobe. It is fused superiorly with the tail of the caudate nucleus and the putamen. The most ventral portion of the clostrum also fuses with the amygdaloid nucleus.

The amygdala is located beneath the uncus of the hippocampus and is usually divided in a series of nuclei: superior, medial, and a basolateral group. The anterior portion of the amygdaloid complex receives the lateral olfactory tract. In the basolateral nuclear group the posterior pilar of the fornix ends. This nucleus also connects with the orbital aspect of the frontal lobe. Stimulation of the superior and medial portion of the amygdala produces pupillary changes and increasing blood pressure and gastric motililty. A lesion in the temporal lobe can produce temporal lobe epilepsy manifested by:

1. Eye movements.
2. Uncinate crisis, manifested by a perception of a foul smell or perfume. More often than not, the patient complains that he or she smells burning rubber.
3. Gustatory hallucinations, everything might taste sweet, sour, or bitter for the patient.
4. He or she might present with smacking of the lips, masticatory or gustatory movement, guttural noises, and/or hyperventilation.

To remember all these disorders, I invite the lector to think in the "midline." Remember that the optic chiasm is in the midline and then comes the nose, and then the mouth, larynx, heart, stomach, and genitalia. Some of the visual hallucinations that the patient might present with are due to the presence of the Meyer's loop that ends at the calcarine fissure; so realistically, when a mass in the temporal lobe produces visual hallucinations, these are simply distal manifestations of proximal stimulation. Before elaborating further about the temporal lobe, let me give an example of a common fact. When a person, for example a man, sees a beautiful woman, he recognizes that she is beautiful; it is called cognition. He wishes to date her and finally makes the decision to invite her for dinner (*conaction*). She agrees, but her beauty is so great that his pulse becomes accelerated, his blood pressure might increase, and his hands are shaking (physical changes). Coming back to the subject of temporal lobe seizures, let us therefore, say that the patient could have visual hallucination manifested by one special characteristic. The objects of hallucination, like seeing a person who seems alive but changes shape and form, is what is known as *metamorphosis*. The person might look like he is extremely small (*Lilliputism*). A patient who was an alcoholic left his wife and the wife died soon after. While he was drinking, he "saw his wife" and began running in a desperate fashion. He fell and developed a *subdural hematoma* that was diagnosed by means of a cerebral arteriogram. He underwent surgery. The next day, and for several days after, he was constantly saying "I don't want to see my wife, I don't want to see my wife." The house staff interpreted this as *delirium tremens*. In spite of the sedation, he was constantly repeating that he did not want to see his wife. After the arteriogram was rereviewed, it was found that the patient had an *arteriovenous malformation* at the tip of the temporal lobe. The patient was obviously in *status epilepticus* and agreed to a second surgery so that the hallucinations of seeing his wife would stop, as indeed it happened. Coming down in the midline, we find the nose; therefore, we can have the previously described olfactory hallucination. Then we find the mouth, and we will find the hallucination as described before, gustatory and masticatory, as well as the smacking of the lips. Continuing in the downward direction, we find the larynx and hence, the hyperventilation and guttural noises. When I mentioned in the previous example about the person that sees a beautiful woman, I simply gave this example to speak about *cognition*. Some patients have the sensation that they are familiar with the place they are in (even though they have never been there), this is known as *déjà vu*. Or on the contrary, if they are in a place that should be well known, they feel they have never been there, *jamais vu*.

A *temporal lobe seizure* can produce tachycardia, bradycardia (irregular heart beat), a sensation of hunger, abdominal cramps, erection without ejaculation, ejaculation with erection, wet dreams, and frigidity. Also in these patients, one can find auditory or musicogenic seizures. Patients may hear voices when no one is around.

Psychomotor seizures or the epileptic equivalent are also found in temporal lobe seizures. The patient might go from one place to another and will not have any recollection of how he or she got there. Psychomotor seizures are characterized, most of the time, by purposeful motor activity that happens without the awareness of the patient, and these are usually followed by an *amnesia* of the event.

It must be also stated that not all the psychomotor seizures originate in the temporal lobe. Due to the connection in between the frontal and the temporal lobe, on certain occasions, a *frontal lobe* lesion might produce a *temporal lobe* seizure.

Lesions of the *hippocampus* can affect *recent* memory. Bilateral ablation of the temporal lobe tip can produce *Kluver-Bucy syndrome*, manifested in the monkey by an inability to recognize their natural enemies. They have oral tendencies (put everything in their mouths). They examine everything they get in their hands with their mouth. They have exaggerated sexual activity and a voracious appetite.

Temporal lobe connection: In a human, the *medial geniculate nucleus* projects the impulse from the *lateral* lemniscus to the auditory cortex. This fiber passes through the sublenticular portion of the internal capsule and then changes direction inferiorly to the inferior border of the insula to reach the *auditory* cortex.

Efferent fibers from the auditory cortex extend to the medial geniculate nucleus and to the inferior colliculus. There are also association fibers that connect the acoustic areas with the most inferior portion of the postcentral convolution (sensory cortex) and with the inferior aspect of the frontal lobe. These fibers, which extend to the frontal lobe, are an integral part of the inferior fronto-occipital bundle. Some of these fibers extend to the cortical center that controls eye movements. There are extensive connections in between the auditory cortex with the insula. The vestibular system also has connection with the temporal lobe cortex at the second temporal convolution. Within the temporal lobe itself, there are numerous association fibers that interconnect the different gyri. The temporal lobes of the two hemispheres are widely interconnected. Rostrally, the middle and inferior temporal convolutions are interconnected via the anterior white commissure. The remaining portions of the temporal lobes are interconnected via the corpus callosum. We have already discussed the interconnection of the temporal lobe and auditory cortex with the visual cortex. Due to the fact that the visual radiation passes *laterally* (from the lateral geniculate body) and curves posteriorly around the trigone of the lateral ventricle, it has been found that lesions (like infarcts that affect the most posterior and medial portion of the temporal lobe) can produce a *homonymous* visual field defect.

Insula   7

The *insula*, or *island of Reil*, lies deep in the sylvian fissure. It can only be seen if the operculi is opened, inferiorly and superiorly, in a widely separated fashion. The *insula* has the shape of a mountain, with the base toward the deep white matter. It is surrounded by a deep sulcus known as the limiting sulcus—which separates the insula from the frontal lobe and the parietal and temporal lobes. The apex of the insula, or tip of the mountain, is directed forward and inward toward the anterior perforated space; and it has wide continuity anteriorly with the most posterior portion of the orbital aspect of the frontal lobe. The insula is divided by the sulcus centralis, or central fissure of the insula, into a *precentral* and a *postcentral* lobe. This sulcus begins at the tip of the insula and extends back and upwards toward the limiting sulcus. The precentral gyrus portion is further divided by shallow sulci into three small gyri, the precentral gyrus and three or four gyri brevis. Immediately behind the central fissure of the island, the postcentral lobe or *gyrus longi*, which often bifurcate at the upper part, can be seen.

The functional significance of the cortex of the insula is as follows: the precentral gyrus of the insula is concerned with movement of the jaw, tongue, and throat, which corresponds also to the frontal operculum of the sylvian fissure. Stimulation at this level produces mastication and salivary movement. Stimulation of the gyrus longi, as well as the stimulation of the parietal operculum of the sylvian fissure, creates the sensation of taste in the tongue, throat and mouth.

The insula overlies the corpus striatum and is separated from them by the *clostrum*, and by the very thin *capsula extrema* and *capsula externa*. At the present, with the exception of some speech changes that do not always occur, removal of the insular cortex does not produce neurologic deficits. Stimulation of the insula produces strange epigastric sensations with feeling like belching and increased gastric mobility. It should be remembered that over the insula lie many branches of the *middle* cerebral artery. The capsula extrema is a thin band of fibers that is situated in between the clostrum and the cortex of the insula. Some authors consider the clostrum as a deep portion of the cortex of the insula. Posteriorly, one can see that there is a fusion in between the clostrum and the cortex of the insula. In the capsula extrema exists some fibers that connect the temporal and frontal lobes. In general, very little is known about all the connections of the insula. It is believed that supplementary motor areas for the face also exist in this area.

The Limbic System   8

The limbic system includes the cerebral cortex of the cingulate gyrus that wraps around the corpus callosum and continues posteriorly with the subcallosal gyrus, gyrus fornicatus, the hippocampus gyrus, dentate fascia, the amygdala, and the hippocampus. The limbic lobe projects many of its axons to different parts of the brain, to the amygdaloid nucleus, the anterior nucleus of the thalamus, the basal ganglia, the hypothalamus, and the pons. The limbic system is concerned with the *autoregulation of visceral organs*. Stimulation of this system produces pupillary changes, lacrimation, and changes in blood pressure, respiration, urination, and defecation. The intralimbic system is formed by the olfactory tract and the two subdivisions, medial and lateral. As stated previously, the medial branch of the olfactory tract wraps around the corpus callosum, and is called the *tract of Lancisi*. It joins the one on the opposite side, above the body of the corpus callosum, forming the *induseum grisseum*, and continues posteriorly with the name of *fasciola cinerea*, *fascia dentata*, and *tract of Giacomini*, which ends at the cortex of the uncus of the hippocampus.

The *fornix* is an integral part of the intralimbic system. The fornix has a triangular shape with its base posteriorly and its vertex anteriorly (see Figure 8-1). It rests in the most medial portion of the superior aspect of the thalamus. The tela choroidia is interposed between the third ventricle and the fornix. The posterior portion of the fornix fuses to the anterior aspect of the splenium of the corpus callosum. Beneath the splenium of the corpus callosum, it forms the posterior columns of the fornix, which extends downward at the sulcus of the hippocampus all the way to the amygdala nucleus with the name of *fimbria*. There are fibers that come from one hippocampus to the other forming the hippocampal commissure. These fibers, together with the pillar of the fornix, have the shape of a *lyra*, known as *corpus psalterium*, or *David's Lyra*. It can, therefore, be said that the fornix forms an arch-shaped structure of white color just beneath the corpus callosum. The fornix is divided in two halves by the septum pellucidum. Some authors consider this as two different portions that join in the midline. Anteriorly, each half of the fornix has one

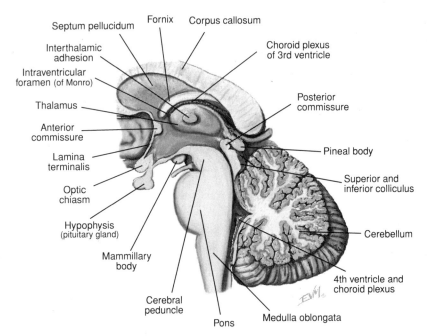

Figure 8-1 A median sagittal view of the brain stem and cerebellum.

pillar; therefore, there are two anterior pillars or columns. The anterior pillars arch downward and backward toward the hypothalamus separating themselves from each other on their downward course, leaving a very narrow space in between. They descend through the gray matter of the hypothalamus, lateral wall of the third ventricle, and behind the anterior white commissure to reach the mammillary body. Each pillar curves around itself to form a loop around the mammillary (*corpus album*). From the mammillary body, we can find a column that climbs upward in the gray matter of the lateral wall of the third ventricle. This is called the mammillothalamic bundle, or bundle of Vicq d'Azyr. This bundle is an integral part of the interlimbic system. In *gross dissection*, it does not appear that the anterior pillar of the fornix and the mammillothalamic bundle have a direct communication.

Between the anterior pillars of the fornix and the anterior thalamus there is an opening on each side known as the *foramen of Monro*. It communicates the lateral ventricle to the third ventricle.

Many fibers of the fornix do not end in the mammillary body. After these fibers decussate, they extend to the mesencephalon, which has been described as extending to the ocular motor nucleus.

As a part of the limbic system, there is a bundle that extends from the interpeduncular nucleus to the nucleus of the habenula—known as the *retroflex bundle of Meynert*.

HIPPOCAMPUS

The hippocampal gyrus is an area located in the temporal lobe. It continues anteriorly, with the name of uncus of the hippocampus, and posteriorly, it is continuous with the callosomarginal gyrus. Above the superior border of the hippocampus is the fascia dentata, the Ammon's horn, and the end portion of the *lamina terminalis*. The two hippocampi are joined together by the hippocampal commissure, which is located between the two posterior pillars of the fornix, forming what was already described as *corpus psalterium*. The main portion of the hippocampus ends at the ventral portion of the splenium of the corpus callosum.

Ammon's horn and its cortex has seven layers.

1. The ependyma
2. The alveus
3. The stratum oriens
4. The double pyramidal layers
5. The stratum radiatum
6. The stratum lacunosum
7. The stratum molecularis

The white matter that covers the ventricular side of Ammon's horn is the *alveus*. In the alveus, there are efferent fibers originating from the axon of the double pyramidal layers. These axons join the fimbria and the fascia dentata. The double pyramidal layer of Ammon's horn is very characteristic of it. This layer is also called *Rose's field*. It is extremely vulnerable to anoxia at birth. Between Ammon's horn and the hippocampal gyrus there are several areas. The subiculum is found medially, toward the ventricular system. It continues with the prosubiculum and laterally, the subiculum continues with the presubiculum. The presubiculum laterally and superiorly intertwines with the entorhinal cortex. In the upper portion of Ammon's horn the fimbria can be identified as it ends in the cortex of the hippocampus. Medial to the *fimbria*, into the temporal horn of the lateral ventricle, one can identify the *choroid plexus*.

The Basal Ganglia   9

The basal ganglia, or basal nucleus, of the telencephalon are located at the base of the brain. The *basal ganglia* refer to a conglomeration of large gray matter masses located at the base of the brain. There is a general consensus that the corpus striatum, putamen, and cauda nucleus; and the globus pallidus, the amygdala, and the clostrum should be the structures to be considered when discussing the **basal ganglia**.

In some books, there is a considerable amount of interrogation as to whether or not the thalamus should be incorporated based upon the large interconnection that exists between these structures.

The term *corpus striatum* is applied to both the *lenticular* and *caudate nuclei*. At times, the word neostriatum is applied to these two structures. The name of *paleostriatum* is one that is applied, by many authors, to indicate this with the *globus pallidum*. The caudate nucleus has the shape of a comma sign. It is a mass of gray matter that has three portions: the head, the body, and the tail. The dorsal part of this large mass is seen in the frontal horn of the lateral ventricle. It is covered by a glistening layer of the lateral ventricle, the *ependyma*. The head of the caudate nucleus pushes inward the anterior horn of the lateral ventricle. It is located anterolateral to the thalamus. It extends posteriorly, decreasing in diameter, and becomes a band that is known as the tail of the caudate nucleus. This curves around the thalamus to end in the amygdaloid nucleus, located at the temporal pole.

The caudate nucleus fuses with the putamen and becomes thicker at the most inner portion; this part is known as *nucleus accumbens septi*. The tail is located at the lateral portion of the body of the lateral ventricle—at the most dorsal and lateral portion of the thalamus, curving down, backward, and finally forward. It is located in the temporal horn of the lateral ventricle, at the most medial and superior portion. Along its course, it is accompanied by an ill-defined tract, the *stria terminalis*—which also ends at the amygdala. The head of the caudate nucleus, at the most ventral and anterior portion, is in anatomic continuation with the putamen, forming a common mass. In some areas, they are incompletely separated by the anterior limb of the internal capsule. The nucleus lenticularis can only be seen in coronal, or transaxial, portion of the brain. When dissection is made horizontally, it has the appearance of a biconvex lens, but in the coronal section, it has a triangular shape with its base located superiorly and externally. The apex of the nucleus lenticularis pushes the internal capsule inward creating the *genu* of the internal capsule. A vertical section of the lenticular nucleus at the midportion allows us to identify two white bands, which are parallel to each other and divide this mass in three zones. The most lateral part is known as *putamen*; the medial part is known as *globus intermedius*; and inner part is known as *globus pallidus*. One can see, in a section, a large number of radiated fibers. The histologic appearance of the masses present multipolar cells, both small and large cells.

In the medial border of the putamen, one can identify the white band known as the *lateral medullary lamina*. The outer surface of the lenticular nucleus is separated from the clostrum by a very thin white band known as the *external capsule*. From medially to laterally, one can see the external capsule, the clostrum, and capsula extrema, and then the insula. It should be stated, however, that there is not a sharp line of demarcation between the ventral portion of the putamen and the amygdala. The *internal capsule*, which separates the thalamus from the lenticular nucleus, is crossed by a large number of fibers that interconnect the thalamus with the lenticular and caudate nuclei, as well as with other lower areas of the upper brain stem. The internal capsule is a large conglomeration of fibers that extends from the cortex to the cerebral peduncle. It has the shape of an opened fan, with the base of the fan directed toward the cortex. The internal capsule is divided into an anterior limb, which is located between the head of the caudate nucleus and the lenticular nucleus; and a genu or knee, which is limited laterally by the apex of the lenticular nucleus. The *posterior limb* of the internal capsule is located in between the lenticular nucleus; and laterally and medially by the tail of the caudate nucleus and dorsal thalamus. There is a segment known as the sublenticular portion of the internal capsule that extends laterally just beneath the lenticular nucleus towards the temporal lobe. Also, there is a retrolenticular portion of the internal capsule that contains fibers that subsequently fan out toward the white matter of the cerebral hemisphere.

There are extensive connections between the lenticular and caudate nucleus with the dorsal thalamus and the motor cortex of the brain. Also, there are connections with the brain stem. It can be stated that they are corticostriate fibers, thalamostriate as well as striatothalamic fibers. The connection between the corpus striatum and substantia nigra and between the globus pallidus and hypothalamus are pathways that are rather complex but very important in the understanding of *movement disorders*. The cortex is connected with the corpus striatum via collateral fibers from the corticospinal tract, and there are also connections of the putamen and pallidum with the motor and premotor cortex. As stated previously, there are wide connections between the thalamus and the corpus striatum, both afferent and efferent.

There are connections between the ventral thalamus, the centrum medianum, and the lenticular nuclei. From the globus pallidum, there is a large bundle of fibers that runs across the posterior limb of the internal capsule to the *zona incerta*. It is known as the *fasciculus thalamicus*. This connects the globus pallidus with the ventral oralis, anterior nucleus of the thalamus, and probably, with the ventrolateral nucleus of the thalamus. Connection in between the substancia nigra and the corpus striatum are *afferent* connections between the basal ganglia. This particular bundle of fibers goes across the globus pallidus to reach the putamen. It is rather large. A lesion in this particular bundle can produce tremors and other movement disorders. There is a large amount of *efferent* fibers that contain pallidosubthalamic fibers. The fasciculus lenticularis, the ansa lenticularis, and subthalamic fasciculus form a conglomeration of fibers that interlace with the internal capsule. The hypothalamic segmental bundle that joins the hypothalamus with the globus pallidus also extends to the cerebral cortex, which controls the movement and expression of the face. It is an important connection for the facial expression of emotions. This particular tract is the one that could explain, in part, the expressionless face observed in a Parkinsonian patient.

The ansa lenticularis originates at the putamen, globus pallidus, and globus intermedius, and receives fibers from the premotor areas, the caudate nucleus, forming a fine anatomic bundle that locates at the sublenticular portion of the internal capsule. This bundle runs in the medial portion of the lenticular nucleus and extends backward and medially in front of the posterior limb of the internal capsule to reach the nuclei of the *field of Forel*. It then extends to the red nucleus.

In spite of the fact that many of the initial primitive movements of the fetus in the uterus are probably controlled through the basal ganglia (i.e., grasping), a lesion of the caudate nucleus may produce little effect. If confined to the head of the caudate nucleus, NO evidence of motor deficit or movement is found; however, if the lesion is *bilateral*, the patient could have difficulties in grasping. When dysfunction at the basal ganglia occurs, two major abnormalities could take place: disturbance in the tone and a disorder in the automatic movements. For example, there may be an inability to swing the arm or abnormal movements like tremor, choreiform disorder, or athetoid movement. These abnormal movements can occur at slow or fast motion. This disorder can also appear independent of a basal ganglia lesion by a destructive process that affects the cerebral cortex, the cerebellum, the thalamus, and substantia nigra. Some of these movements occur at rest, like in Parkinson's syndrome. In Parkinson's syndrome, the abnormal movements do NOT occur during sleep. The tremor of Parkinson's disappears also with intentional movement. In Parkinson's syndrome, there is a widespread degeneration of the globus pallidus, the substantia nigra, and the dorsal thalamus. The hypertonicity in the gait can occur in syndromes that effect the lenticular nucleus.

The Thalamus 10

The optic thalami are two large masses located on each side of the lateral ventricles medially, and the corpus striatum laterally. It has six aspects: superior, inferior, medial, lateral, anterior, and posterior. The dorsal thalamus occupies the major portion of the diencephalon. For the most part, there are many descriptions of the nuclei of the thalamus that often create confusion for the reader. The thalamus rests over the cerebral peduncle. The nuclei of the thalamus can be divided into:

1. The nucleus of the midline, whose function is primarily visceral, and widely connected with the hypothalamus.
2. The anterior group, which is connected to the frontal lobe and the limbic system.
3. The medial group, which has wide connection with the hypothalamus.

It can also be said that the thalamus is divided by a white band into *internal* and *external* masses, the *intra-* and *extra-laminary nuclei*. The lateral mass contains another thin white band of fiber called the external medullary lamina. The internal medullary lamina has a **Y** shape. The nucleus located outside of the lateral medullary lamina is known as the *external reticular* nucleus of the thalamus. Medial to the intramedullary lamina, we find the intralaminar, or medial, group. This medial mass is subdivided into medial nuclei and intralaminar nuclei. In general, the accepted description of the thalamus is as follows (see Figure 10-1).

1. A portion known as metathalamus, which is made from the medial and lateral geniculate nuclei.
2. The lateral nuclear group, which has two portions, a ventral group and a dorsal group.
3. The thalamic reticular nucleus.
4. The anterior nuclear group.
5. The posterior nuclear group.
6. The intralaminary nucleus, which contains two portions, the medial nuclei and the nucleus medialis (of the intralaminary nucleus).

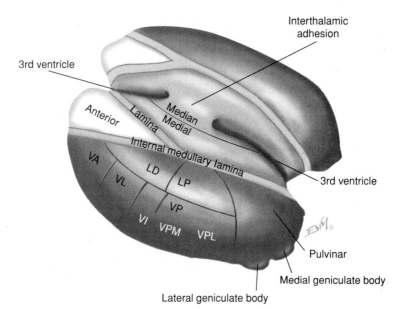

Figure 10-1 A schematic of the thalamus.

The anterior nuclear group has three portions, or subnuclei: the anteromedial, the anteroventral, and the anterodorsal. These nuclei have wide connections with the hypothalamus and mammillothalamic bundle, and with the amygdala and singular gyrus. The reticular nucleus of the thalamus is a conglomeration of nuclei located between the external medullary lamina and the internal capsule. Ventrally, they extend to the zona incerta. In the case of cerebral infarct that destroys the most anterior portion of the frontal cortex, there has been found degeneration of the most anterior group of the reticular nucleus, which indicates that there are wide connections with the cortex of the frontal lobe. The intralaminary group is the conglomeration of a nucleus located medial to the medial laminary lamina. It can be subdivided into two portions, one more medial and one more lateral.

The *medial group* has four nuclei:
1. Nucleus medialis dorsalis
2. Nucleus submedialis
3. Nucleus medialis ventralis
4. Nucleus parafascicularis

The *intralaminary nuclei* are those closely attached to the internal medullary lamina and there are also four nuclei:
1. Centrum medianum
2. Nucleus centralis medialis
3. Nucleus parafascicularis medialis
4. Nucleus lateralis centralis, also called *semilunar nucleus of Dejerine*

This group of nuclei has a wide connection with the hypothalamus and with the cerebral cortex. It appears, however, that the centrum medianum does not have a cortical connection. However, when there is lesion in the insular cortex, some cells of the centrum medianum degenerate. The centrum medianum has wide connection with the putamen and the caudate nucleus. The overall function of the centrum medianum is not known. The nucleus dorsum medialis, which is located medial to the internal medullary lamina, continues forward for a few millimeters and posteriorly with the pulvinar. This nucleus receives fibers from the nucleus of the habenula. A lesion in this nucleus can produce cardiovascular and respiratory disturbances. The dorsal medialis nucleus also has wide connections with the orbital cortex by way of the anterior thalamic radiation. Its connections are considered very important for emotional expression. Bilateral destruction of the dorsomedial nucleus produces a flat affect—similar to the changes observed in patients with *frontal lobectomy*. The patients also have an incorrect concept of time, for example, they think that the day will be over within two hours—even though it is early morning.

In the posterior part of the medial nucleus, the nucleus parafascicularis is located along with another one, not mentioned before, known as *nucleus parataenialis*.

It appears that this nucleus receives fibers from the spinothalamic tract, although its connections are not well established.

LATERAL NUCLEAR MASS

This mass can be subdivided into a dorsal portion and ventral portion, pulvinar, pars angularis, and nucleus reticularis. The dorsal part is divided into the nucleus *lateralis dorsalis* and nucleus *lateralis posterior*. Some authors clump the nucleus lateralis dorsalis, the nucleus lateralis posterior, and the nucleus ventricularis together in one group. The ventral thalamus is subdivided into:

1. Nucleus ventralis anterior (VOA).
2. Nucleus ventralis lateralis.
3. The posterior ventral thalamus, which has two nuclei, the VPL, nucleus ventralis posterolateralis; and the VPM, nucleus ventroposteromedialis.

The VOA has wide connection with the putamen and the pallidum. The VL receives the largest cerebellar outflow via the superior cerebellar peduncle and connects with the premotor cortex. A lesion in the posterior ventral group of the thalamus can produce the following abnormalities. In the VPL, ending in the medial lemniscus, one can find *contralateral* loss of the vibratory sense, position sense, and light touch (remember the three letters VPL: V for vibratory sense, P for position sense, and L for light touch). The tactile function of the face may not be affected because of its bilateral projection of the V nerve. Sometimes, the patient has hyperesthesia. Considering the VPM nucleus (which receives the contralateral spinothalamic tract), if the patient has a lesion that is "irritative," the patient can present with an unbearable *burning pain*. This pain can be provoked by the lightest contact with a sheet or by blowing air into the skin (because of neural degeneration of the nuclei).

These three findings of: contralateral hemi-tremor, contralateral hemi-ataxia, contralateral loss of pain and temperature; plus, contralateral loss of touch, vibratory sense, position sense; plus, the intractable pain are known as the syndrome of *Dejerine Roussy*. The ventral thalamus contains the zona incerta and the nucleus, or field of Forel. The zona incerta extends into the whole length of the cerebral peduncle, just ventral to the thalamus. It contains a large number of small neurons, fibers that arise from the globus pallidum, the lenticular fasciculus, and fibers from the fasciculus subthalamicus.

The pulvinar is located in the posterior portion of the thalamus and has wide projection to the parietal and temporal lobe. In the pulvinar, there are functions related to visual association. This is an intermediate nucleus that receives impulses that subsequently *relay* to the cerebral cortex.

The posterior nuclei group, as stated before, has:

1. The lateral geniculate body
2. The medial geniculate body.
3. The suprageniculate nucleus.
4. The nucleus limatens.

The geniculate bodies are very important. The medial one receives auditory impulses that subsequently are transferred to the first and second temporal gyri. The lateral geniculate body receives visual impulses from the optic tract, and subsequently sends them to the occipital cortex.

In the subthalamic region along the medial border of the internal capsule, we find the subthalamic nucleus of Luys. The subthalamic nucleus is connected with the dorsal thalamus, the field of Forel, the zona incerta, the red nucleus, and the substantia nigra. A lesion of the subthalamic nucleus produces violent ballistic movements known as hemiballismus.

Both thalami are connected at the midline by a gray commissure known as masa intermedia.

EPITHALAMUS

The epithalamus is the region that involves the posterior white commissure, the pineal gland, the habenula, and the habenular commissure. The posterior white commissure is a conglomeration of crossing fibers. However, a large group of cells are an integral part of it. These cells organize in a clump from the dorsal nucleus of the posterior commissure. Ventral to it, there is another nucleus called the nucleus of *Darkschewitsch*. The fibers of the posterior white commissure interconnect the pretectal nucleus. From the nucleus of Darkschewitsch, also called the *interstitial nucleus of Cajal*, originate direct and cross fibers that form part of the medial longitudinal fasciculus.

The *pineal gland* is located behind the third ventricle and rests in the quadrigeminal plate. It develops from the posterior part of the diencephalic roof. The gland reaches the maximum definitive size at the age of 5 years and it may have a length of about 10 mm and a width of 3 to 5 mm. This gland is connected and fixed to the nucleus of the habenula and to the caudal commissure by glial tissue also called peduncles of the pineal gland. Inside the pineal gland, there is a cavity known as the *pineal recess* of the third ventricle. The nerve fibers that enter the pineal gland originate from the posterior white commissure. The pineal gland has a hormone known as *melatonin* (acetyl-5-methoxytryptamine). This gland produces a very little amount of melatonin. This hormone induces drowsiness and in the hen, stimulates nesting. This hormone also increases with darkness. If this gland grows to be a tumor, it can produce clinical manifestations due to pressure over the aqueduct of Sylvius, vein of Galen, and quadrigeminal plate, producing hydrocephalus and/or precocious puberty. If it compresses the superior cerebellar peduncle, it can produce bilateral *ataxia*. After the age of 15 years, calcifications are found in the pineal gland. The functions of the pineal gland are not as well known as the dysfunction that occurs following a tumor.

THE HABENULA

The habenula, one for each hemisphere, has a medial and a lateral nucleus and it continues medially with the one on the other side, forming the habenular commissure. The most important connections of each habenula are the habenulotectal tract, the habenulotegmental tract, and the habenulopeduncular tract. The habenulopeduncular tract connects the habenula with the interpeduncular nucleus and is also known as the retroflex bundle of Meynert. Another important connection of the habenula is the stria medullaris of each hemisphere. The stria medullaris extends forward at the medial border of the thalamus and descends forward behind the foramen of Monro to the anterior hypothalamus. This portion is known as the *lateral preopticohabenular tract*.

The habenula also has a long thin anterior connection that runs above the dorsomedial nucleus of the thalamus and descends to the most medial group of the preoptic nucleus of the hypothalamus. This connection is known as the *habenular tract*.

The geniculate bodies are divided into medial and lateral structures.

The medial geniculate nucleus is the main part of the area known as the metathalamus. Each medial geniculate body appears as a tubercle at the anterior end of the superior colliculus, inferior to the pulvinar. Posteriorly, it receives a large amount of fibers from the lateral lemniscus. Each medial geniculate body has a wide connection of fibers toward the auditory cortex at the superior and medial temporal gyri. It is considered an important station of the auditory pathway with the cerebral cortex at the temporal lobe.

The *lateral* geniculate nucleus is located inferolaterally in the pulvinar. It has a rather pear-like shape. Anteriorly, it receives the *optic* tract. In coronal sections, it has the appearance of a *Napoleon cap* and it has six laminas, or layers. Between the different layers of the lateral geniculate body, the multiple bands of optic tract fibers can be found. In the posterior portion of the lateral geniculate body, one can find the optic radiations, which end at the occipital cortex.

The Hypothalamus and Pituitary Gland   11

The hypothalamus occupies the most ventral portion of the diencephalon and develops from the alar portion of the diencephalon. It can be stated that it is the area that integrates the visceral activities, endocrine glands, and smooth muscles. In sagittal planes, the hypothalamus is divided in two halves, right and left. Superiorly, the hypothalamus is demarcated from the thalamus by the hypothalamic sulcus. Anteriorly, the hypothalamus overlaps and interacts with the preoptic area. When we see the ventral aspect of the cerebral hemisphere, the hypothalamic area is outlined frontally by the optic chiasm, laterally by the optic tract, and posteriorly by the two cerebral peduncles. In this mesial aspect of the hypothalamus, from anterior to posterior, one can see, in the center, the tuber cinereum or central eminence, the infundibular stalk with the hypophyseal recess of the third ventricle, the two mammillary bodies, and the area of the posterior perforated space.

The hypothalamus weighs about 4 grams and it extends from the lamina terminalis and anterior white commissure to a plane that goes to the level of the posterior white commissure. The hypothalamus can be divided into the anterior, medial, and posterior hypothalamus. The anterior hypothalamus has several nuclei.

1. The supraoptic nucleus that straddles over the optic chiasm. It contains a large number of neurosecretory cells and is highly vascular. The fibers of the neurons of these nuclei descend through the pituitary stalk to the posterior hypophysis, or neurohypophysis, and from the stalk, it continues to the so-called *pars nervosa* of the pituitary gland. These nuclei are primarily concerned with water metabolism.
2. The two **p**reoptic nuclei are located immediately below the anterior white commissure. Stimulation of the preoptic nucleus promotes the secretion of oxytocin, which is necessary to fix the concept of love, which is created when the frontal lobes are bathed with dopamine. In animals, the stimuation produces **p**inpoint pupils.

3. The paraventricular nucleus has two portions, a medial and a lateral. The lateral part of the paraventricular nucleus is also known as the nucleus filiformis. Stimulation of the filiform nucleus elicits *rage*. A lesion of either one of these two nuclei, medial or lateral portions, can produce a very peculiar abnormality. In this abnormality, the eyes present with seesaw eye movements.

The anterior hypothalamic area extends through the supraoptic portion and intertwines with the axons of the most anterior part of the middle hypothalamic area. These nuclei also control the temperature, and it could be called the thermostat of the body. It is a hypothalamic regulator of temperature. The supraoptic and paraventricular nuclei are located in the most lateral portion of the optic chiasm.

Diabetes insipidus and adipsia is produced when there is a deficiency of the ADH, antidiuretic hormone, by a lesion in the *supraopticohypophyseal system*. Diabetes insipidus has also been described in lesions of the paraventricular nucleus. The antidiuretic hormones preserve the body water by reabsorbing the glomerular filtrate at the tubulous area of the kidney. This hormone travels from the supraoptic nucleus into the neuroapophysis by way of the supraopticohypophysial tract. This hormone is very important in the control of serum *osmolality*. Deficiency of this hormone produces diabetes insipidus, manifested by *compulsive water drinking*.

The *medial* hypothalamus has two portions: a central and a lateral part (see Figure 11-1). The central eminence, or *tuber cinereum*, is subdivided into a dorsal and ventral portion. These are the dorsomedial and ventromedial nuclei of the medial portion of the medial hypothalamus, and they are formed

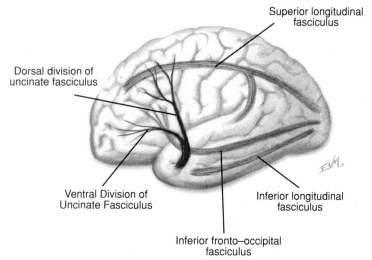

Figure 11-1 The fasciculi.

by a conglomeration of autonomic cells. The lateral portion, or lateral hypothalamic system or mammilloinfundibular nucleus, is connected to the orbital portion of the frontal lobe via the forebrain bundle. Destruction of the ventromedial nucleus leads to *hyperphagia*. In the ventromedial nucleus, there is a center for appetite and a satiety center. Certain patients can present with a hypothalamic dysfunction characterized by hyperphagia, polydipsia, and hypersomnia. The syndrome is called *Kline-Levine syndrome*, and often, it is associated with restlessness; patient could feel persecuted. Students may want to remember "Calvin Klein" underwear and the "Levine" tube to remember the association with this syndrome.

The ventromedial nucleus has neuronal control of carbohydrate metabolism. A lesion of the ventromedial nucleus can produce *hyperosmolar diabetic nonketogenic coma*. Great attention should be given to this syndrome in patients that have had hypothalamic surgery, because it can be accompanied by an acute demyelination at the level of **C7**, with the catastrophic result of *irreversible quadriplegia*. It is also well known that patients that undergo surgery of the hypothalamus in the area of the tuber cinereum, may have hyperacidity and ulceration of the stomach.

Stimulation of the ventral and dorsomedial nuclei of the hypothalamus produces secretions that are conveyed by the axons of the neurosecretory cells to the pituitary stalk, and from the pituitary stalk to the anterior lobe of the pituitary gland. In the pituitary stalk, there is a conglomeration of arteries and veins (this is called the portal system of veins and arteries) that enters the anterior pituitary lobe. Among the hormones elicited by the neural hormones from the dorsomedial and ventromedial nucleus, is the *ACTH*, adrenocorticotrophic hormone. This hormone stimulates the adrenosympathetic system for fight or flight.

Examples of the trophic hormones are follicule stimulating hormone, luteinizing hormone, THRH, prolactin, and ACTH.

Stimulation of the most anterior portion of the medial hypothalamus can produce shivering or hyperhydrosis (sweating). Crisis of rage can be produced by lesions that involve the dorsal or ventromedial nuclei of the hypothalamus. Lesions at these two nuclei can also produce *precocious puberty*, which is also found in tumors of the pineal gland that invade the dorsomedial nucleus of the hypothalamus.

The posterior hypothalamus contains the mammillary bodies and large nuclei known as the posterior hypothalamic nucleus. This nucleus has axons that descend to various portions of the upper brain stem.

MAMMILLARY BODIES

The *mammillary bodies* are rounded masses located behind the tuber cinereum rostral to the posterior perforated space. The *mammillary bodies* are divided into a medial portion, which is greatly affected by *alcoholism* (e.g., *Wernicke-Korsakoff syndrome*). This nucleus has two distinguished portions within it. The lateral part is rather small and is also called the nucleus lateralis of the mammillary body. This nucleus is sometime called the *nucleus intercalatus*. The mammillary body has wide connection with the anterior nucleus of the thalamus and with the fornix. In its course through the hypothalamus, the fornix is surrounded by a small group of neurons that contain large cells; this is the *perifornical nucleus*. The hypothalamus is connected with the most dorsal portion of the brain stem, particularly to gustatory impulses that are incorporated in the medial lemniscus that ends in the VPL nucleus of the thalamus, and from there to the central eminence of the tuber cinereum. As mentioned previously, the mammillary bodies connect with tegmentum of the brain stem via mammillotegmental fibers.

Neurons of the olfactory system that end in the hypothalamus project their axons to the preoptic nucleus of the hypothalamus.

MEDIAL FOREBRAIN BUNDLE

The *medial forebrain bundle* is a conglomeration of ascending and descending fibers that connect the osmatic center (for smell) of the cerebral hemisphere with the hypothalamus and the mid-brain. The area of the hypothalamus that is interconnected with the paraolfactory area is the ventromedial nucleus of the hypothalamus; therefore, it can be said that the middle forebrain bundle connects the orbital aspect of the frontal lobe with the medial hypothalamus.

The majority of the preoptic nucleus and the anterior hypothalamus in general, are connected with the *habenula* through fibers that come from the habenula anteriorly, and laterally to the preoptic region. There is a conglomeration of fibers that originate from the amygdaloid nucleus that follow backward along the tail, body, and head of the caudate nucleus, and subsequently curve downward and are divided into fibers that are supra- and precommissural (above and anterior to the anterior white commissure). It also has subcommissural fibers that reach the preoptic area of the hypothalamus, connecting, in this fashion, the amygdala with the hippocampus. There are also fibers that connect the dorsomedial nucleus of the hypothalamus through the anterior peduncle of the thalamus to the frontal cortex. There are abundant connections between the interpeduncular nucleus and the mammillary bodies.

HYPOPHYSIS

The *hypophysis* is a gland that originates from the *Rathke's pouch* (an embryonic diverticulum that comes from the nasopharynx and enters the sella turcica). The hypophysis forms the *pars distalis*, the *pars intermedius*, and the *pars tuberalis* of the gland. These three parts are known as the *adenohypophysis*. The portion that comes from the infundibular system of the hypothalamus is known as the *pars nervosa*. It is composed of a type of cell known as pituicytes, and nerve fibers that descend from the tuber cinereum, and the stalk, from the supraoptic nucleus and periventricular nuclei. The hypophysis is found in the sella turcica, covered on top by an incomplete diaphragm that has an opening, which the pituitary stalk goes through. This gland has a blood supply from the superior hypophysial arteries, branches of the intracaverous portion of the carotid artery just as the artery exits the cavernous sinus. This artery runs medially in the pituitary stalk. There is a specific branch, the inferior hypophysial artery, that originates from the cavernous portion of the carotid artery and is divided in medial and lateral branches. This artery nourishes the *pars nervosa*. From the pars nervosa, antidiuretic hormone and oxytocin have been isolated.

The *anterior* lobe of the pituitary gland produces hormones that have a profound affect on the metabolic and endocrine function of the body. These hormones are the following.

1. Growth hormone (GH)
2. Thyroid-stimulating hormone (TSH)
3. Adrenocorticotropic hormone (ACTH)
4. Follicule stimulating hormone (FSH)
5. Luteinizing hormone (LH)
6. Prolactin hormone (PH)

The removal of the anterior pituitary gland produces a panhypopituitarism manifested by hypofunction of the thyroid gland, diminished production of the corticosteroids, and gonadal failure. When you look at the pituitary gland from below (through the nose), consider the secretions in tumors to be associated with the following areas.

1. Anterior and lateral-prolactin secretion (prolactinoma)
2. Posterior and lateral-ACTH
3. Midline-TSH
4. Posteriorly and midline-GH (gigantism)

Hyperproduction of growth hormone produces *acromegaly, prognathism*, thickened fingers and toes, *macroglosia*, prominent nose, thick lips, and a massive chest. If the hyperproduction occurs in a *child, gigantism* will occur. *Gigantism* is overgrowth with relatively normal proportion. In cases of *acromegaly*, autopsy reveals generalized hypertrophy of the viscera, heart, stomach, intestines, and liver. This condition occurs with an increase in the number of eosinophilic cells.

Chromophobe adenomas are the ones that produce the largest growth of the pituitary gland. Chromophobe adenomas occur more often in the third or fourth decade of life. These tumors tend to produce compression of the adjacent or neighboring structures, involve the optic chiasm, and give rise by *temporal hemianopsia*. It should be remembered that the prolactin-secreting tumors are the most common tumors of the pituitary gland in women. In a man with a "dry" ejaculation, but normal erection, consider a *prolactinoma* as a cause because seminal vesicle contraction is inhibited by the excess prolactin.

The basophilic adenomas are usually small tumors and they are associated with so-called *Cushing's syndrome*, which is characterized by buffalo neck, facial acne, and skin striaes. In women, it is accompanied by amenorrhea. The majority of these patients have arterial hypertension, hyperglycemia, and elevated ACTH. There is a common tumor of the pituitary gland known as a *craniopharyngioma*. This occurs primarily in *children*, and arises from the embryonic epithelial rest of the Rathke's pouch. These tumors are usually large cystic masses and are highly calcified. It could also be fleshy in characteristic. When the tumor is large, it can produce blindness and optic atrophy by compression of the optic chiasm. The tumor can also compress the third ventricle, producing hydrocephalus by blocking the drainage of the cerebrospinal fluid from the third ventricle to the aqueduct of Sylvius. These cysts contain a large quantity of dark brown fluid with abundant cholesterol crystals. If a person has a pituitary tumor that produces ACTH and the surgeon by mistake removes the adrenal glands, then the feedback mechanism to the adrenal gland disappears. A new syndrome originates, *Nelson syndrome*, and the pituitary grows tremendously. In a woman, the nipple is hyperpigmented and the person looks as if she has a constant tan. If the surgeon accidentally removes the *adrenal gland* and recognizes the mistake, then the surgeon has to TOTALLY remove the pituitary gland. If not, the tumor of the pituitary gland in Nelson's Syndrome will continue to recur after every tumor-removal surgery.

There is a syndrome known as *adiposogenital dystrophy (Frohlich's syndrome)*. It is characterized by a tumor that involves *both* the hypophysis and the hypothalamus. There is a great deal of retardation of the sexual development, and increased appetite. The adiposity does not occur in lesions that involve the pituitary gland *alone*. Postpartum hemorrhage can also produce necrosis of the anterior pituitary gland leading to *panhypopituitarism*, known as *Sheehan's syndrome*.

The Corpus Callosum 12

The corpus callosum (CC) is a thick stratum of transverse nerve fibers that connects every part of one hemisphere to the other hemisphere. The corpus callosum forms the largest white commissure of the brain. The CC forms the roof of the frontal portion and trigone of the lateral ventricle. From its origin, known as the rostrum of the corpus callosum—when seen in the medial aspect of the brain—it has an anterior vertical portion that has a sharp bend and becomes thicker, receiving the name of the *genu* of the corpus callosum. It extends horizontally from one hemisphere to another as it runs to the occipital lobe, continuing with a rounded end known as the *splenium* of the corpus callosum. The entire CC has a length of about 10 to 11 cm.

Again, the anterior portion of the corpus callosum is known as a beak, or *rostrum*. The *splenium* of the corpus callosum curves forward and is then continued by the *fornix*. As mentioned before, in the body of the corpus callosum runs the *tract of Lancisi*, also known as *stria longitudinalis*. It has an anatomic relation with the anterior cerebral arteries and the singular gyrus. The part of the corpus callosum that curves *forward* from the genu into the frontal lobe in each hemisphere, is known as the *forceps minor*, or *forceps anterior*. The part that curves *backward* from each side of the splenium in the occipital lobe, is known as the *forceps major*, or *forceps posterior*. The body of the corpus callosum extends all the way lateral to the temporal lobe after it has covered and formed the roof of the lateral ventricle—also known as *tapatum*.

Sometimes, the corpus callosum might be totally absent, or in part. A patient with this may lack any signs, or it may be associated with idiokinetic apraxia in the left side. A lesion of the corpus callosum is usually accompanied by involvement of the *cingulate gyrus* (e.g., in Marchiaffava-Bignani's disease—which is a degeneration of the corpus callosum occurring in people who drink homemade red wine). In the inferior aspect of the corpus callosum, there is a glistening surface, the *ependyma*. In the midportion of the corpus callosum, we have the *septum pellucidum*, which divides the two lateral ventricles into left and right. The septum pellucidum extends from the fornix below to the corpus callosum above. It has the shape of a virgula, with a posterior portion that is very narrow.

Among the other great commissural fibers, is the *hippocampal commissure* that extends from one hippocampal gyrus to another. These fibers course through the *posterior* pillar of the *fornix*, or fimbria, running beneath the splenium of the corpus callosum, joining in this fashion, the two hippocampi. The *anterior white commissure* is a bundle of white fibers that runs in front of the *anterior* pillar of the fornix. The main bundle of fibers runs transversely. It has several components.

1. The inter-amygdaloid and inter-hippocampal component
2. Fibers that connect one corpus pallidum with the other
3. The stria terminalis

There is a wide communication between the middle, inferior, and superior temporal lobes. These form the largest component of the anterior white commissure.

In the white matter, there are a large number of *association fibers*.

1. The fasciculus of the cingulus that lies in the medial part of the hemisphere just above the corpus callosum. Rostrally, it extends around the genu of the corpus callosum to the paraolfactory area. Posteriorly, it curves around the splenium of the corpus callosum, and follows a downward and forward course within the hippocampal gyrus to the uncus of the hippocampus.
2. The uncinate fasciculus, previously described with the description of the temporal lobe, connects the temporal lobe to the frontal lobe. It has two divisions, a *ventral* one (that extends to the orbital part of the frontal lobe), and a *dorsal* portion (that extends to the most anterior part of the mid-frontal gyrus).
3. The inferior longitudinal fasciculus is a large group of white fibers that extends from the tip of the temporal lobe to the lingual gyrus of the occipital lobe.
4. The inferior occipitofrontal fasciculus has a direct connection in between the occipital lobe and the frontal lobe, and occupies the most inferior part of the extreme capsule in a very intimate relation with the uncinate fasciculus.

The *occipitofrontal fasciculus* associates the frontal pole with the visual cortex. It lies below the corpus callosum, above and laterally to the head of the caudate nucleus.

The *fasciculus arcuatus*, also known as the *superior longitudinal fasciculus*, is located at the dorsolateral border of the putamen, above the claustrum, and in close relation to the internal capsule.

The *posterior commissure* is a rounded band of fibers that extends from one optic thalamus to another. Many of the posterior commissure's fibers connect the anterior and superior lamina quadrigemina of one side with the one on the other side. It also contains fibers that originate in the pineal body and the habenula.

The *internal capsule* is formed by fibers that descend from the cortex to the brain stem, plus fibers from the corpus striatum and optic thalamus. In its horizontal section, we can find two portions.

1. The *anterior limb* of the internal capsule, which separates the *lenticular nucleus* from the *caudate nucleus*.
2. The *posterior limb*, which separates the *optic nucleus* of the thalamus from the *lenticular nucleus*.

The midportion, where the anterior and posterior limbs join, is known as the *genu*. It is pushed in medially by the globus pallidus. The internal capsule has large fibers that arise from the cerebral cortex of the cerebral hemisphere. The fibers of the anterior limb of the internal capsule arise from the frontal lobe cortex. The one from the genu and anterior two thirds of the posterior limb originate from the rolandic area of the cortex. The posterior third of the posterior limb comes from the temporo-occipital region. Besides the previously described fibers, there is a large conglomeration of fibers from the cortex. These fibers terminate in the thalamus and corpus striatum. In a coronal section, it has the appearance of a fan and is known as the *corona radiata*. In transaxial views, it has an oval shape and is known as the *oval center of Fleshing*. This extends downward into the cerebral peduncle. (A further description will be given once the description of the brain stem is reviewed.)

The Ventricular System   13

The ventricular system is divided into the lateral ventricles, third ventricle, aqueduct of Sylvius and fourth ventricle (see Figure 13-1).

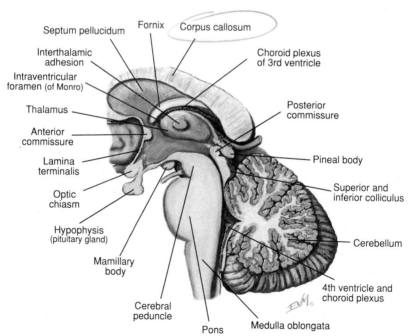

Figure 13-1   A median sagittal view of the brain stem and cerebellum.

LATERAL VENTRICLES

The *lateral ventricles*, one for each hemisphere, have a frontal horn, body, trigone of the lateral ventricles, occipital horn, and temporal horn. The lateral ventricles are separated from each other by the septum pellucidum. They have a wide communication through the foramen of Monro with the third ventricle. The lateral ventricles are covered by a thin and shiny membrane (known as the *ependyma*). The lateral ventricles have a roof, a floor, and an inner wall. The roof is formed by the inferior surface of the corpus callosum. The floor is formed by: the caudate nucleus, the optic thalamic nucleus, the tinea semicircularis, the choroid plexus, and the fornix (posterior pillar).

The *anterior* horn of the lateral ventricle extends from the foramen of Monro into a downward slope and into the frontal lobe, curving laterally over the head of the caudate nucleus. It is limited medially by the septum pellucidum and laterally by the junction of the corpus callosum, with the corpus striatum. The *body* of the lateral ventricle is located below the lower part of the parietal lobe. Its roof is also formed by the *corpus callosum*, and its inner wall by the septum pellucidum. The choroid plexus and the fornix form the *floor* of the body of the lateral ventricle. The *posterior* horn occupies the occipital lobe. It curves backward and outward. The roof is also formed by fibers of the corpus callosum. The inner wall has a longitudinal eminence, known as *calcar avis*. The calcar avis is the intraventricular representation of the calcarine fissure.

Joining the temporal horn is the *trigone* of the lateral ventricle or confluency of the frontal, occipital, and temporal horns.

The *temporal* horn of the lateral ventricle extends from the trigone into the temporal lobe of the brain inferiorly and laterally to the optic thalamus; all the way anterior to the tip of the temporal lobe. The roof is formed primarily by the inferior aspect of the tapatum, or the tail of the caudate nucleus, and the taenia semicircularis. The floor has several portions: the hippocampus major, the pes hippocampi, the collateral eminence, the fimbria, and the choroid plexus. In the mesial aspect, there is found the *transverse fissure* through which the pia matter invaginates into the ventricle to cover the choroid plexus.

The *tela choroidea* is a modified layer of the ependyma. It extends from one inner border of the thalamus to the one on the opposite side. It lies below the fornix. The tela choroidea contains the *choroid plexus* of the third ventricle and the two *internal cerebral veins*, posteriorly. The tela choroidea continues with the arachnoid and is also known as *velum interpositum*. The choroid plexus curves around the foramen of Monro and runs backwards extending to the trigone of the lateral ventricle where it enlarges; forming the *glomus* of the choroid plexus, and from there extends to the temporal horn.

THIRD VENTRICLE

In the roof of the third ventricle, there is a telencephalic structure that originates at the junction of the choroid plexus with the ependyma as an embryonic structure that subsequently involutes—it is known as the *paraphysis*. No physiologic importance or neurologic value is found in humans. However, if it does *not* involute, it can change into a tumoral structure known as a *colloid cyst* of the third ventricle. The colloid cyst of the third ventricle can block both foramina of Monro and/or the third ventricle. By and large, the colloid cysts have a long pedicle. One of the classical clinical presentations is that the person may become unconscious when they bend forward, and when he or she wakes, they have a violent *headache* due to the *acute hydrocephalus* that occurs secondary to the obstruction of the foramen of Monro. If the person tilts the head backwards and shakes the head, he or she may have another syncopal episode; but once the patient comes to, the headache is gone. In personal experience, I have operated on 62 cases of colloid cysts of the third ventricle and only one had this history.

The lateral ventricles are intercommunicated with the slit-like cavity known as the *third ventricle* via the foramen of Monro. The foramen of Monro has its superior limits by the anterior pillar of the fornix, and the inferior part by the anterior nucleus of the thalamus.

The third ventricle is located in between the two thalami and is separated from the thalamus by the hypothalamic sulcus. The floor of the third ventricle is formed by the hypothalamus. Its anterior wall is formed by the lamina terminalis and the anterior nucleus of the hypothalamus. The *roof* of the third ventricle is formed by the *tela choroidea* and above the tela choroidea by the *fornix*. Posteriorly, the walls are formed by the posterior white commissure and the pineal gland. There is a recess in the pineal gland itself, and another integrated by the arachnoid above the pineal gland, known as the suprapineal recess. In the third ventricle, there is a gray commissure known as *masa intermedia*.

AQUEDUCT OF SYLVIUS AND FOURTH VENTRICLE

The aqueduct of Sylvius communicates the third with the fourth ventricle. The fourth ventricle communicates with the subarachnoid space by the two foramina of **Luschka**, laterally, and by the foramen of **Magendie**, medially. This structure will be described in detail with the description of the brain stem.

CEREBROSPINAL FLUID

Cerebrospinal fluid (CSF) is a crystalline fluid discovered by Cotugno, and is located inside the ventricular system and circulates in the *subarachnoid* space. It is also found in some dilated portions of the subarachnoid space known as *cisterns*, the *cisterna magna* being the largest one. The cerebrospinal fluid is *formed* in the *choroid plexus* and the *ependymal* cells by filtration and secretion. CSF is also *absorbed* by the ependymal cells and the subarachnoid villi. The normal appearance of this liquid is crystalline. It normally contains 10 to 30 mg of protein, 40 to 80 mg of glucose, and 7 to 10 mg of chloride, and has a pH of 7.45. It also contains several amino acids, leucine, and thyroxine. In certain pathologic diseases, the variation of these normal components can be observed. Normally, the cerebrospinal fluid contains about 3 to 4 cells per every cubic millimeter. The majority of these cells are *lymphocytes*. If the cells increase, it could be said that there is *pleocytosis*. The increase of these cells is observed as a reaction of the meninges in an infectious or inflammatory process.

When the pleocytosis exists in an inflammatory process and/or infectious process, the cells are usually *polymorphonuclear cells*. In an acute infectious process, the proteins are elevated and the glucose is decreased. In certain illnesses, like Guillain-Barré syndrome, there is *albuminocytolytic dissociation*. The albumin is *elevated* and the cells are almost not present or very low. Alteration of the glucose of the cerebrospinal fluid is found in several entities. In acute meningitis, as well as in tubercular meningitis, there is a profound *decrease* in the blood sugar. In an infection with a *viral* process, like encephalitis, there is a relative *increase* of the cerebrospinal fluid glucose. The chlorides of the cerebrospinal fluid are primarily found in tubercular meningitis.

Studies of cerebrospinal fluid have an enormous value in several entities. These can be obtained by cisternal puncture or by lumbar puncture. One could study biologic reaction in cases of syphilis, increase in the protein in cases of tumor, infectious processes, and degenerative diseases of the brain such as demyelinating diseases. The cerebrospinal fluid pressure varies between 50 and 180 mm of water, that should be measured with the patient in a lateral and horizontal position. An increase of the cerebrospinal fluid is found as an increase of the intracranial *pressure*. Of great importance is the presence of CSF that is found in intracranial bleeding due to a ruptured aneurysm, intraparenchymal bleed, rupture into the ventricles, arteriovenous malformation, and certain metastatic tumors (like a melanoma). Lumbar puncture is *contraindicated* in the presence of increased intracranial pressure! It is strongly recommended that before a lumbar puncture is done, that a CAT scan or MRI is done. Spinal puncture should NOT be done in the presence of intracranial hematomas, masses, or tumors.

In many cases, a *headache* occurs following a lumbar puncture. To avoid such headaches, the patient must be maintained in a prone position.

HYDROCEPHALUS

Hydrocephalus is a condition characterized by *excessive* cerebrospinal fluid within the ventricular system. It can be divided into external and internal hydrocephalus. Some authors describe hydrocephalus as obstructive and nonobstructive, or communicating or non-communicating—implying that an obstructive process exists between the lateral and third ventricle, third ventricle and aqueduct, or fourth ventricle and subarachnoid space.

Internal hydrocephalus is the case when excessive cerebrospinal fluid *within* the ventricular system is produced by an illness or process that interferes with the *circulation* of this fluid. *External hydrocephalus*, in contrast, is *secondary* to a process *outside* of the ventricular system. The sites of obstruction are usually: the foramen of Monro, the third ventricle, the aqueduct of Sylvius, the fourth ventricle, and the foramina of Magendie and Luschka.

Unilateral occlusion of one foramen of Monro produces ipsilateral dilatation of the ventricle where the occlusion is present.

If the patient is a female in her first decade of life, one should think of an *ependymoma* or a *choroid plexus papilloma*. Obstruction at the level of the third ventricle can be intrinsic or extrinsic. Intrinsically, we can find an ependymoma of the third ventricle, colloid cyst of the third ventricle, gliomas, and meningiomas. Extrinsically, we can find tumors of the pineal gland, a large tumor of the pituitary gland, a large aneurysm of the basilar or anterior communicating artery, large meningiomas of the tuberculum sellae, and/or gliomas of the optic chiasm. At the aqueduct of Sylvius, pathology can be extrinsic *or* intrinsic.

Extrinsically, one could find a tumor of the pineal gland and an aneurysm of the vein of Galen or *meningioma* of the tentoria incisura. Intrinsically, the most common problem is a *stenosis* of the aqueduct of Sylvius —likely produced by an inflammatory reaction and/or ependymomas. A pathologic condition at this level is known as *forking of the aqueduct*, and is highly associated with *Arnold-Chiari malformation* (medulla oblongata and the cerebellar tonsils protrude into the spinal cord area —blocking the CSF flow at the 4th ventricle). Forking of the aqueduct is commonly associated with a *myelomeningocele*.

The occlusion at the fourth ventricle could be due to a process that is within the fourth ventricle itself, such as ependymomas, neurocysticercosis, or choroid plexus papillomas. Of great interest is that one of the most common causes of gastric outlet obstruction in an infant is tumors of the floor of the fourth ventricle. The extrinsic occlusion could be due to a tumor of the *cerebellar hemisphere* and/or a tumor of the *vermis* of the *cerebellum*.

Occlusion of the foramina of Luschka and Magendie can be congenital. This produces a large cystic dilatation of the fourth ventricle. It is accompanied by absence of the inferior vermis of the cerebellum and the posterior inferior cerebellar artery. In this condition, the ventricular system does not communicate with the subarachnoid space. The tentorium and the lateral sinuses are very elevated, and the straight sinus follows an almost vertical course. These patients can present with a profound ataxia. Due to the pressure that is exercised over the brain stem, the patient can have a *twining voice*, difficulty swallowing, and, of course, hydrocephalus.

External hydrocephalus, or *communicating hydrocephalus*, is a term that is reserved for instances when the etiology of the hydrocephalus lies *outside* the ventricular system. This may be associated with: occlusion of the arachnoid villi (like in meningitis), subarachnoid hemorrhage, cortical vein thrombosis, occlusion of the lateral sinuses, and an excessive amount of protein in the subarachnoid space (like in cases of Guillain-Barré syndrome; proteinaceous fluid).

The symptomatology and clinical manifestations of *hydrocephalus* vary between children and adults.

In an *infant*, in whom the sutures are still open, the *head* becomes *enlarged*, the anterior fontanelle becomes full and tense, the veins of the scalp become engorged, and the skin is tightly stretched over the skull bone. The eyes are deviated *downward*. The patient is unable to look upward and may develop optic disk atrophy. Paralysis of cranial nerve *VI* could also occur and the child may present with marked *spasticity*.

In children that already have a *closed* fontanelle, complaints include headache and a presentation of papilledema, nystagmus, and/or Parinaud syndrome (paralysis of the upward gaze and lack of convergency). These children can also present with spasticity of the lower extremities. In the adult, the patient complains of headaches that may or may not be accompanied by emesis, papilledema, nerve VI palsy, and paralysis of the upward gaze.

In certain circumstances, the adult might present with a syndrome known as *normal or low pressure hydrocephalus*, which is characterized by the following:

1. Dementia
2. Ataxia
3. Urinary sphincter incontinence
4. Impaired memory
5. Spastic gait

The hydrocephalus is also accompanied by other spectific manifestations, depending on the site of the obstructive process, for example: *panhypopituitarism*, as in a tumor of the pituitary gland; or hypothalamic syndromes, if the hydrocephalus is due to a tumor of the hypothalamus. If there is a *colloid cyst* of the third ventricle, besides the already described manifestations, the patient could have loss of consciousness with changes in position, impaired memory, and seesaw eye movements. In a tumor of the pineal gland, the hydrocephalus can also be accompanied by precocious puberty or testicular and/or ovarian atrophy.

In congenital occlusion of the foramina of Luschka and Mangendie (Dandy-Walker sydrome), the patient can present with (in addition to the hydrocephalus) profound cerebellar ataxia, nystagmus, twining voice, and difficulty in swallowing. Currently, treatments of hydrocephalus are multiple. Scarff and Stokey, in occlusion of the aqueduct of Sylvius, have performed a third ventriculostomy and/or perforation of the lamina terminali. Multiple operations, known as *shunts*, are also done: ventriculoperitoneal shunt, ventriculopleural shunt, ventriculoatrial shunt, and shunting of the fluid into the ureter. I have done more than 80 ventriculo-gallbladder shunts. In occlusion of the aqueduct of Sylvius, the lateral ventricles are connected to the cisterna magna. This procedure is known as ventriculocisternostomy or Torkildsen procedure. Others have removed the choroid plexus. From time to time, hydrocephalus can spontaneously arrest, possibly due to transependymal absorption of the cerebrospinal fluid. There is not a clear statistic of the incidence of spontaneous arrest. In cases of cerebral atrophy, the ventricular system becomes enlarged and the term for this condition is known as *hydrocephalus ex vacuo*.

The Midbrain 14

The midbrain is also known as the *mesencephalon*. It is a portion of the human brain that is located at the base of the brain between the brain and the pons. The mesencephalon comprises the quadrigeminal plate, the aqueduct of Sylvius, the geniculate bodies, and the cerebral peduncle. It has an inclined plane from the thalamus down and backwards. Above, the midbrain continues with the brain and below, with the pons. It has three surfaces, one ventral, one dorsal, and two laterals. Some authors describe only a ventrolateral and a dorsal portion. In the ventrolateral surface, the cerebral peduncle appears as a very short, stalk-like, cylindrical white column surrounded by the gyrus of the hippocampus and by the posterior cerebral arteries and basal vein of Rosenthal. Most of its fibers are vertical. In the most ventral and medial portion, is found the *interpeduncular space*, which has a *triangular* shape. At the interpeduncular space, one can identify cranial nerve III (the oculomotor nerve) as it emerges from the cerebral peduncle.

The dorsal aspect of the *cerebral peduncle* can only be observed after separating the two cerebral hemispheres. It presents four eminences; two superior and two inferior—separated by a cruciform sulcus. These structures are known as *lamina* or *corpora quadrigemina*, or simply the *quadrigeminal plate*. In the most lateral part of the quadrigeminal plate one can identify a vertical, somewhat forward-inclined sulcus that separates the cerebral peduncle itself from the quadrigeminal plate. To study the different anatomic structures located within the cerebral peduncle, it is necessary to do several cross-sections of the peduncle. In the transaxial section, one can see that each cerebral peduncle is divided by a very *pigmented* layer known as *substantia nigra*, which divides the cerebral peduncle in two portions—one anterior portion (known as *pes peduncularis*) and the dorsal portion (known as *tegmentum*). The pes peduncularis of each peduncle is separated one from the other by the interpeduncular space and can be divided in three portions.

The outer third portion brings fibers that course through the posterior limb of the internal capsule. Some of these fibers originate in the pons and extend to the cerebral cortex of the temporal and occipital lobe. The three medial thirds contain fibers that take origin at the motor cortex and descend from there into the internal capsule, genu, and posterior limb—which, in the pons and medulla oblongata, receive the name of *pyramid*.

The *substantia nigra* is made of a conglomeration of neurons that contain a *melanin* pigment. It extends through the entire length of the peduncle from the thalamus to the pons. It contains three portions: the *reticulated* portion, *pars lateralis*, and the *area compacta*. The pars lateralis, as the name indicates, is located in the most lateral part of the substantia nigra. This portion has a wide connection with the putamen, pallidum, and the thalamus. It also connects with the tectum. The reticulated area has a bundle of fibers that extends to the corpus striatum, intertwining with fibers of the internal capsule. Some of the fibers also connect with the ventralis oralis anterior (VOA) nucleus of the thalamus and from there, to the frontal cortex. It is believed, as a result of many studies, that the *area reticulata* (of the substantia nigra) has an important function in the *stabilization* of *voluntary movements*.

Posteriorly, the substantia nigra has crossed and uncrossed fibers that connect the substantia nigra with the *red nucleus*. The substantia nigra is largely affected in patients with postencephalitic *Parkinson's*.

Behind and medial to the substantia nigra appears the red nucleus. It is called red nucleus due to its reddish appearance. It is *encapsulated* by fibers of the superior cerebellar peduncle (on their way to the VL nucleus of the thalamus) and lies below the optic thalamus. This capsule is known as the *white nucleus*. The red nucleus has a cylindrical shape and measures about 1.3 cm in height, with a diameter of about 4 mm. It is composed of large and small groups of cells. The *large* group of neurons is known as the *magnocellular nuclei* and the *small* group of cells is known as *parvocellular group*.

The magnocellular portion receives the great cerebellar outflow proceeding from the *dentate* nucleus. These fibers are decussated in the brachia conjunctivus. There is a large afferent corticorubral fiber that originates in the cerebral cortex of the frontal lobe. There are also contralateral and homolateral fibers that are from the superior colliculus.

The *lenticular nucleus* is connected with the red nucleus via *fasciculus lenticularis*. There are large, efferent descendent fibers (known as the *rubrospinal* tract) which connect the red nucleus with the nucleus of the facial nerve and trigeminal nerve. The red nucleus provides muscle tone and orientation in space. A lesion of the red nucleus can also produce *palatal myoclonus*.

Conglomerated around the red nucleus, particularly more towards its caudal portion, there is a series of *tegmental gray nuclei* among which run innumerable fibers (in their passage from the lower brain stem to the thalamus). Adjacent to the periaqueductal gray matter, is a group of coalescent nuclei known as *cuneiform nucleus*. All of the cells of these tegmental gray nuclei receive synopsis from the *reticular system*, from the spinal cord, and the lower brain stem. Their function is probably *visceral*. In this myriad of nuclei, is the end the hypothalamotegmental tract (which also connects to the red nucleus).

Several small nuclei are located between the ventrotegmental area and the hypothalamus, more specifically with the mammillary bodies. Lesions in the *ventrotegmental area* are associated with *somnolence*. Some optic fibers, as they go across the vertical fibers of the cerebral peduncle, extend toward a group of small nuclei lateral to the substantia nigra and are known as *tractus peduncularis transversus*.

PERIAQUEDUCTAL GRAY MATTER

The periaqueductal gray matter is located lateral to the aqueduct of Sylvius and represents the origin of the *oculomotor cranial nerve III*. Each cranial nerve III receives fibers from the ipsilateral as well as the contralateral nuclei. The nucleus of cranial nerve III has a common central area known as the *Edinger-Westphal* nucleus (centrally located). Anterior and lateral, it is ventrocaudally the nucleus for *accommodation*, behind and inferior to the previous nucleus is the *photomotor center*. Laterally to these previously described centers, from superior to inferior, is the center for the elevator palpebral, the superior rectus, the internal rectus, the inferior oblique, and the inferior rectus. When lesions occur at the nucleus of origin of cranial nerve III, partial external ophthalmoplegia can occur, meaning that the pupillary sphincter is spared. At times, a type of ophthalmoplegia known as *progressive nuclear ophthalmoplegia* can be observed. This is manifested first with *ptosis*, followed by *paralysis* of other *extrinsic muscles* of the ocular globe. Cranial nerve III then crosses forward through the medial portion of the red nucleus to exit in the interpeduncular space. This nucleus overlies the medial longitudinal fasciculus and fuses to the one at the opposite side at the inferior pole, an area that is known as the *central nucleus of Perlia*.

The Edinger-Westphal nuclei receive visual impulses that are relayed to the superior colliculi by the tecto-oculomotor fibers. The Edinger-Westphal nuclei send impulses via cranial nerve III to the *ciliary ganglia*. Therefore, in lesions produced by *syphilis*, there can be loss of the light reflex, with preservation of accommodation. This is known as *Argyll-Robertson pupils*.

Lateral to the periaqueductal gray matter, one can identify the mesencephalic root and nucleus of the trigeminal nerve. This root can be followed by using a special stain down to the masticatory nucleus, representing the proprioception of mastication. In the midline, ventral to the periaqueductal gray matter, one can identify a conglomeration of cells forming the nucleus of Raffe, just below a plane crossing through the inferior collicular levels.

There is a very important bundle of parallel periventricular fibers, which are an important and integral part of the dorsal longitudinal fasciculus. This fasciculus carries impulses from the hypothalamus and tectal region to all the periaqueductal nuclei.

The tegmentum of the cerebral peduncle is an extremely important area through which the pathways go from the spinal cord and lower brain stem to the thalamus and to the cerebral cortex.

THE MEDIAL LEMNISCUS

The proprioceptive impulses traverse the posterior spinal ganglion and the nerve root and enter the spinal cord through radicular fibers that integrate the posterior column in the spinal cord, with the name of the bundle of Gall and Burdach. Then, all the way up to the medulla oblongata—it establishes synopsis with the cellular body of the nuclei *cuneatus* and *gracilis*. The cuneatus and gracilis nuclei form a very well-identified structure, known as the *clava*, at the inferior pole of the fourth ventricle. As they ascend lateral and anterior to the fourth ventricle, some of the fibers begin to decussate. They reach the pons, where they decussate at the middle third, forming the *medial lemniscus*. They ascend behind the pyramidal tract curving upward and laterally to occupy an area located lateral to the red nucleus and dorsal to the substantia nigra, being capped laterally by the medial geniculate nucleus, to reach the ventroposterolateral nucleus of the thalamus. From there, through the posterior limb of the internal capsule, they reach the sensory strip of the parietal cortex. Posterior to the medial lemniscus is the deep tegmental gray matter.

The fibers that bring *tactile* or *superficial sensitivity* after they leave the peripheral nerve also reach the spinal ganglion of the posterior root. They establish synopsis at this level. As they enter the spinal cord, they are divided into two groups of fibers, *long* and *short* fibers, establishing synopsis with the cells of the posterior column.

The *short* fibers, which carry *pain* and *temperature*, enter the posterior column—where they establish synopsis, decussate two segments above, crossing the midline, and locate anterolaterally in the spinal cord. This forms the *dorsolateral spinothalamic tract* of the *spinal cord*.

The *long* fibers are those that carry the *tactile* impulses. They enter the spinal cord at the level of the posterior column. At three segments superiorly establish synopsis with another sensitive neuron, which continues its ascending course through another two or three segments, to finally locate at the most *anterolateral* portion of the spinal cord of the *opposite* side, integrating in this fashion, the *ventral spinothalamic tract*.

They continue in an ascending course until this tract reaches the medulla oblongata. This *ventral spinothalamic tract* traverses the reticular formation of the medulla oblongata where it establishes synopsis and joins (at this level) the *dorsolateral spinothalamic tract* ascending through the pons to the cerebral peduncles. At the level of the cerebral peduncle, it joins to the *medial lemniscus* to reach the thalamus. As it traverses the pons, it also joins the quintothalamic tract. For this reason, a lesion located at the plane *below* the emerging root of the *trigeminal* nerve produces anesthesia of the face—at the site of the lesion, with anesthesia of the *contralateral* side of the body (from the neck down).

The *ventral spinothalamic tract*, which carries *pain* and *temperature*, ends in the thalamus, whereas the tactile impulse extends from the thalamus through the posterior limb of the internal capsule to the parietal lobe. In the cerebral peduncle, the ventral spinothalamic tract is located at the most lateral horn of the medial lemniscus, in front of the lateral tectotegmental spinal tract. The ventral ascending fibers of cranial nerve V are located in an intimate location behind the medial lemniscus. The gustatory components are located at the most ventromedial horn of the medial lemniscus.

THE LATERAL LEMNISCUS (VESTIBULAR AND COCHLEAR)

The fibers that integrate the vestibular nerve reach the lateral angle of the fourth ventricle. The internal roots, as they reach the medulla oblongata, are divided in two groups of fibers.

1. The *ascending* fibers, which enter the medulla oblongata, end in three nuclei located at the floor of the fourth ventricle: the nucleus of *Deiters*, the nucleus dorsalis internus, and the nucleus *Bechterew*.
2. The *descending* fibers or inferior roots of the acoustic nerve. The cochlear root or external root enters the restiform body to join the anterior auditory nucleus and the lateral auditory nucleus.

The vestibular root travels transversely and medially, and it decussates in the midline with fibers of the opposite side (at the level of the reticular formation) to become *ascendent* and to join fibers of the acoustic pathway, known as the *lateral lemniscus*. The vestibular fibers have components that run downward following the course of the inferior cerebellar peduncle to end in the nucleus of the roof of the cerebellum. The vestibulospinal fibers have a downward course that ends in the anterior horn cells of the spinal cord. The same vestibular nuclei give origin to ascending fibers, which join the ipsilateral as well as the contralateral medial longitudinal fasciculus, through which they then interconnect with the nuclei of cranial nerves III, IV, and VI. Through this connection, one can explain the *nystagmus* that occurs in damage of the *vestibular system*.

The *cochlear nucleus* gives origin to a group of fibers that travels medially and transversely, traversing to the *superior olivary nucleus*, crossing with the one on the opposite side, and forming the trapezoid body, which is located at the inferior third of the pons. Fibers from the lateral acoustic nucleus travel subependymally to the floor of the fourth ventricle forming the *acoustic* striaes that end in the *superior olivary nucleus* of the pons. The fibers of the *trapezoid body* follow in an upward course, joining in its ascending direction its vestibular fibers, forming the *lateral lemniscus*—which are located in the most lateral portion of the cerebral peduncle to end some of its fibers in the posterior and inferior quadrigeminal tubercles. The main component traverses through the posterior limb of the internal capsule to the superior and medial temporal convolution.

The *medial longitudinal fasciculus* (MLF) is a prominent bundle of association fibers that extends from the upper aqueductal gray matter, are overlapped posteriorly and medially by the nucleus of cranial nerve III, and can be followed as low as the spinal cord. It has a large number of fibers; however, the main components are the internuclear component, extrapyramidal component, and vestibular component from the medial, inferior, and lateral nuclei.

Fibers of this *vestibular* component extend upward to the *nucleus of Darkschewitsch* ipsilaterally as well as contralaterally. Fibers from the superior colliculus, as well as the interstitial nucleus of Cajal, descend in the medial longitudinal fasciculus all the way down to the spinal accessory nerve at the anterior horn cells. This provides fibers to the nucleus of cranial nerves III, IV, and VI. The *medial* vestibular nucleus gives fibers that integrate in the medial longitudinal fasciculus to the ipsilateral and contralateral nucleus of cranial nerves III, IV, and VI. There are also interconnections in the medial longitudinal fasciculus between the nucleus of the facial nerve, cranial nerve V, and cranial nerve III.

The *extrapyramidal* components of the MLF are established through the *nucleus of Darkschewitsh*—since this nucleus has a connection with the *lenticular* nucleus. A lesion in one MLF produces *medial longitudinal fasciculus syndrome* characterized by the following:

1. Weakness in adduction of the eye ipsilateral to the lesion, due to attempts to perform lateral gaze
2. Monocular nystagmus in the opposite abducting eye
3. Normal convergence

The *longitudinal fasciculus* is located ventral to the aqueduct of Sylvius, in the midline (in front of the nucleus of Darkschewitsh).

The *central tegmental bundle*, which extends from the thalamus to the superior olivary nucleus, and the pons has a conglomeration of fibers that bring impulses from the basal ganglia and zona incerta. It is located in close proximity to the periaqueductal gray matter and immediately posterior to the medial longitudinal fasciculus. Clinically, damage of the central tegmental bundle produces a syndrome characterized by *palatal myoclonus*.

At the *cerebral peduncle*, one can find the previously described decussation of the superior cerebellar peduncle, and also the ventrotegmental decussation.

The *decussation* of the cerebellar peduncle lies at a plane going below the inferior colliculi and is located behind the substantia nigra. These fibers *ascend* (some ventrally and some dorsally) to the globus pallidus and to the ventrolateral nucleus of the thalamus.

QUADRIGEMINAL TUBERCLES

The quadrigeminal tubercles are also called *corpora quadrigemina*, and consist of four round eminences (placed: two superior and anterior, and two inferior or posterior), and they are separated by each other by *cruciform sulcus*.

The *anterior tubercles* are the largest of the four and have a pear-shape, with the slender portion of the pear directed forward and outward (the posterior one, in contrast, has a circumferential shape). From the outer and superior portion of each tubercle, left and right, there is a prolongation of white bands that are called the *brachia, or arms*. The *anterior brachia* run outward between the pulvinar and the internal geniculate bodies to end in the external geniculate body.

The *inferior* or *posterior quadrigeminal tubercle* connects to the internal geniculate body via posterior brachia. The outer part of every quadrigeminal tubercle has a white color externally, and it has gray matter interiorly. The white matter that covers the anterior quadrigeminal tubercle is called *stratum*

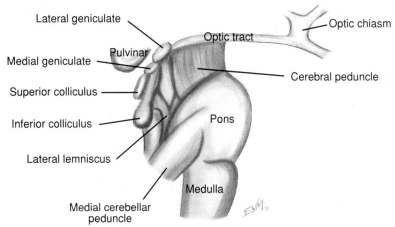

Figure 14-1 Landmarks of the brain stem.

zonale. There is also a *grayish* layer that is known as *stratum lemnisci*, which is interposed between the periaqueductal gray matter of the cerebral peduncle and the quadrigeminal tubercle itself. The inferior cerebellar peduncle literally enters in front of the quadrigeminal plate, and a great portion curves laterally (below the inferior colliculus) to subsequently decussate. The lateral lemniscus, as it approaches the inferior colliculus, spreads around; some fibers enter this nucleus, but the majority of the fibers (of the lateral lemniscus) run dorsolateral to the main nucleus (see Figure 14-1).

Several syndromes can occur at the level of the cerebral peduncle, and they are known as *mesencephalic syndromes* that involve the cranial nerve.

1. **Weber's syndrome.** The base of the midbrain is the site of the lesion and it produces ipsilateral palsy of cranial nerve III, with a contralateral hemiplegia.
2. **Claude's syndrome.** The site of the lesion is at the tegmentum of the pons. As it joins the cerebral peduncle; it causes an *ipsilateral cranial nerve III palsy*, because it also involves the *red nucleus* and the *white nucleus*. It should be remembered that the *white* nucleus is nothing else than the *superior cerebellar penducle,* producing contralateral *cerebellar ataxia* and *tremors* as it wraps around the red nucleus.
3. **Benedikt's syndrome.** The site of lesion is located at the tegmentum of the midbrain and it will involve, of course, the red nucleus and the pyramidal tract, manifesting itself by contralateral cerebellar ataxia and some corticospinal tract signs.
4. **Parinaud's syndrome.** The site of the lesion is at the tectum of the midbrain. The patient will present with paralysis of the upward gaze, fixed pupils, and a lack of convergency.

For further illustration, review Figure 14-2 and Figure 14-3.

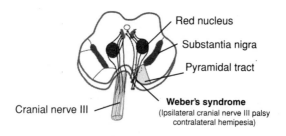

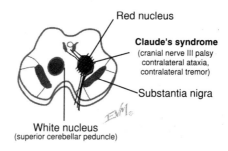

Figure 14-2

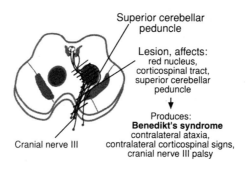

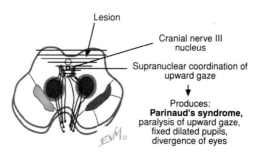

Figure 14-3

The Pons 15

The pons, also known as *pons Varolii*, is an area in which various segments of the encephalon are confluent—connecting through the cerebral peduncle with brain above, with the medulla oblongata below, and posteriorly with the cerebellum. Recall that the **metencephalon** is composed of *both* the *pons* and the *cerebellum*. The *pons* has an anteroposterior diameter of about 2.5 cm, and it is about 3.4 cm in width. It has six surfaces: anterior, posterior, superior, inferior, and a right and left lateral portion. The superior part is in a direct continuation with the cerebral peduncles; the inferior aspect is a direct continuation with the medulla oblongata. The posterior aspect contains the upper part of the fourth ventricle. The ventral surface rests over the clivus and is in direct contact with the basilar artery.

This *ventral* surface has a convex shape and it has a large number of transverse fibers, which extend from one cerebellar hemisphere to the other creating a bridge across the midline adjoining the two cerebellar hemispheres, hence the name of pons, which means bridge. The inferior border of the anterior aspect of the pons forms a well-defined border. Below this border, one can identify the pyramidal tract of the *medulla oblongata*. Exactly in the midline at this particular border, one finds the *foramen cecum*. The cranial nerve VI, right and left, emerges a few millimeters lateral to this foramen. Also in the anterior surface of the pons, there is a large groove produced by the basilar artery. At the junction of the middle cerebellar peduncle with the pons, exactly at the mid-third, is the apparent origin. Better yet, the exit of cranial nerve V from the pons. The *posterior* portion of the pons, as mentioned before, represents the *upper* portion of the *fourth ventricle*. This posterior portion will be described in conjunction with the description of the posterior surface of the medulla oblongata, which in essence represents the inferior triangle of the fourth ventricle.

The pons contains two distinct portions.

1. One anterior or ventral portion (which contains the majority of the transverse fibers that join the two cerebellar hemispheres).

2. A group of *longitudinal fibers* (which represent the different pathways and bundles that ascend from the medulla oblongata to the pons on its way to the cerebral peduncle).

The *dorsal* portion of the pons contains the continuation from the reticular formation of the medulla oblongata and is known as the *tegmental portion*. In the *ventral* portion, besides the superficial transverse fibers and the longitudinal fibers, deep layer of transverse fibers are also found. In reality, there is not a clear demarcation between the superficial and deep fibers. These transverse fibers are nothing else than the *brachium pontis* or middle cerebellar peduncle. The longitudinal fibers of either side are located in each side of the midline. The cluster of cells that are located around the different longitudinal and transverse fibers are known by the name of *gray pontine nucleus*. Most of these cells are rather small or medium in size.

Among the longitudinal fibers, there are large components of bundles that come from the cerebral cortex (particularly fibers from the frontal lobe, which terminate in the more rostral medial pontine gray nucleus). Also, a conglomeration of fibers comes from the temporal lobe known as *Turk Maynert*, which ends at the lateral pontine gray nucleus.

The *dorsal* portion of the pons, as described previously, is the upward continuation of the *reticular formation* of the *medulla oglongata*. It has a medial portion known as the *raphe* and one lateral portion at each side of the midline. The transverse fibers form a very distinct bundle known as the *corpus trapezoideum*, or *trapezoid body*. The fibers incline laterally to connect with the cells of the accessory auditory nucleus. The longitudinal fibers located in these segments are those that proceed from the medulla oblongata. Many of these fibers are located between the trapezoid body and the reticular formation of the pons, representing the upward course of the sensory tract. The most posterior longitudinal fibers, however, contain ascending and descending fibers.

Many of the gray nuclei of the reticular formation represent some of the nuclei of the cranial nerves. Among them, the following should be mentioned.

NUCLEI OF CRANIAL NERVE V

Cranial nerve V, or the *trigeminal* nerve, in the pons, has two nuclei. One represents the *motor* root and the other the *sensory* root. The *motor* nucleus is located in the upper third of the pons, laterally and dorsally to the superior border of the fourth ventricle, just under the ependyma, It is known as *nucleus masticatorious*. The *sensory* nucleus lies somewhat lateral and dorsal to the motor one, just beneath the superior cerebellar peduncle. As it will be seen subsequently, the superior cerebellar peduncles form the lateral boundaries of the pontine portion of the fourth ventricle. The fibers, motor and sensory, run in an anterolateral direction passing through the pons to exit in the ventrolateral portion of the pons. The motor root is well defined and identifiable from the sensory portion of this nerve.

The sensory root of the trigeminal nerve forms a long tract of fibers that represents the ascending root, which can be also be seen in the medulla oblongata and upper part of the spinal cord. The *trigeminal nerve* is the most voluminous of the cranial nerves and, as mentioned before, it has a sensory and a motor branch. The *sensory nucleus* has three nuclei:

1. The first nucleus extends from the medulla oblongata to the pons, and it is known as *nucleus gelatinous of Rolando*.
2. A second, medial nucleus, is the nucleus of the *locus ceruleus*.
3. The third is the *gelatinous nucleus*.

This sensory nucleus extends without interruption as a large column of great substance from the lateral portion of the medulla oblongata and inferior third of the pons. From this sensory nucleus emerges a large number of sensory fibers, which follow an ascending course to finally emerge and integrate the inferior root of the trigeminal nerve.

The *medial nucleus* is located just above the previous one, somewhat behind it. From this nucleus, there is a small bundle of fibers that follow a rather horizontal course in the pons. It is known as the mid-root of cranial nerve V.

Finally, the *nucleus* of the *locus ceruleus* is located in the floor of the fourth ventricle, and from there, a bundle of fibers follow an ascending trajectory integrating what is known as the *locus ceruleus*. These sensory fibers, which will be subsequently described, run toward the petrous bone to reach the cavum Meckel that contain the Gasserian ganglion.

The *gelatinous nucleus* as well as the medial nucleus emits a conglomeration of fibers that runs medially. It crosses in the medial raphe with the fibers of the opposite side, subsequently following an ascending course in the pons and cerebral peduncle and joining the medial lemniscus to finish in the thalamus. During this ascending course, these fibers (of the trigeminal nerve) give origin to collateral fibers, which end in the motor nucleus of the bulboprotuberantial nucleus as well as in the reticular nucleus of the pons. Certain authors believe that the sensory fibers of cranial nerve V are not an integral part of the medial lemniscus, but rather form an independent bundle.

The *masticatory nucleus* receives fibers from the prerolandic cortex of the opposite side. Each cortical center presents bilateral connection with the masticatory nucleus, having in this fashion, a bilateral supranuclear input. There are also connections of the motor nucleus with the globus pallidum. This trigeminal nerve perceives the sensibility of the major portion of the scalp and face, and also is the nerve that controls the masticatory muscle.

NUCLEUS OF CRANIAL NERVE VI

The nucleus of cranial nerve VI is located beneath the ependyma of the pontine portion of the fourth ventricle at each side of the midline. The root of the facial nerve, which has a tortuous course, surrounds medially and posteriorly the nucleus of cranial nerve VI, producing an elevation known as *abducent eminence*. The fibers of the nerve follow an anterior course, traversing through the inferior third of the pons to exit at each side of the *foramen cecum*, at the lower part of the pons and upper portion of the medulla oblongata. This nerve innervates the *lateral rectus muscle*.

NUCLEUS OF THE FACIAL NERVE

The nucleus of the facial nerve has an elongated shape and is located in the pons, dorsal to the *superior olivary nucleus*. The roots that originate from this nucleus, run backward and medially—until they reach the floor of the fourth ventricle. Then, as dissection continues, a rather round bundle that wraps around the nucleus of cranial nerve IV can be found. This forms the previously described *abducent eminence* (immediately after it takes an outward course through the pons). It emerges at the supraolivary fossa between the olivary and restiform bodies of the medulla oblongata. The motor root of the facial nerve is located at the inferior third of the pons. Some of the nuclear cells that give origin to the orbicularis palpebral extend upward to the nucleus of cranial nerve III. While other fibers (in particular, those that innervate the *orbicularis oris*) extend to the nucleus of the hypoglossal nerve.

The facial nerve innervates all the *cutaneous muscles of the face*. In lesions of the lower part of the nucleus of the facial nerve (that extends to the medulla oblongata), a certain number of muscles of the upper portion of the face are *not* paralyzed and are, therefore, spared. Particularly, the *frontalis muscle*, *orbicularis palpebral*, and the *superciliaris muscle* are unaffected. One can say that these muscles are not involved in paralysis of the facial nerve of a cerebral origin. However, these muscles are involved when the lesion occurs in the facial nerve, in its *peripheral* course. One can infer that the nucleus of the facial nerve extends into the medulla oblongata from an "inferior facial nerve," and there is a "superior facial nerve" that controls the upper muscles of the face.

There is a sensory root of the facial nerve, known as the *intermediary nerve of Wrisberg*, which has its origin in the superior portion of the gray ala (nucleus of the fasciculus solitorous). From the nucleus of origin, the intermediary nerve follows an oblique and forward course laterally, also exiting at the supraolivary fossa. From there, the facial nerve extends to the petrous bone to the internal auditory canal where it is accompanied by the auditory nerve. As it will be discussed subsequently, the facial and the intermediary nerve of Wrisberg enter in the Fallopian aqueduct. About 0.5 cm after the facial nerve enters the Fallopian aqueduct, the intermediary nerve of Wrisberg ends at the geniculate ganglion. The geniculate ganglion also sends a branch that intermingles with the facial nerve itself; from the geniculate ganglion on one can say that the facial nerve should be called a mixed nerve. The motor cortical supply is located in the inferior portion of the rolandic gyrus and from there, fibers go through this centrum semiovale to the internal capsule at the level of the genu of the internal capsule. The motor fibers that correspond to the innervation of the superior facial nerve seem to have a different origin than the fibers that go to the genu of the internal capsule, and probably follow the ansa lenticularis.

NUCLEUS OF THE AUDITORY NERVE

The nuclei are two in number, one dorsal, located at the medulla oblongata that extends upward to the pons in the pontine portion of the fourth ventricle; and a ventral nucleus that is partially located in the pons and in the medulla oblongata. This part of the nucleus, in the medulla oblongata, lies anteriorly and laterally to the inferior cerebellar peduncle, between the cochlear and vestibular nerve. The portion located in the pons is located ventral to the restiform body. There is a third nucleus, known as the nucleus of Deiters, located in the most outer portion of the fourth ventricle. The fibers that constitute the vestibular nerve reach the neural axis forming the internal root, and the fibers of the cochlear nerve constitute the external root.

The fibers of the internal root, or vestibular nerve, enter the medulla oblongata at the level of the supraolivary fossa. They run an oblique course backward and are subdivided in two groups of fibers; the ascending one ends in three nuclei located below the floor of the fourth ventricle, known as the nucleus of Deiters, nucleus dorsalis internus, and nucleus dorsalis lateralis. The descending fibers are known as the inferior root of the acoustic nerve. The external root of the cochlear nerve enters the external aspect of the restiform body and ends in two small gray nuclei known as the anterior nucleus of the auditory nerve and the lateral acoustic tubercle.

Several fibers depart from the nucleus of the vestibular nerve:

1. Fibers that run transversely and inward to and decussate in the midline to meet the one on the opposite side and end in a reticular formation. To reach the cerebral cortex, many authors postulate that these fibers follow the same course of the medial lemniscus.
2. *Vestibulo-cerebellar fibers*, which extend from the vestibular nucleus to the cerebellum following the inferior cerebellar peduncle, and end in the nucleus of the roof, in the nucleus globosous and emboliform.
3. *Vestibulo-spinal fibers*. These fibers take origin at the nucleus of Deiters and run to the spinal course with the name of vesibulospinal tract. It is believed that they end at the anterior horn of the spinal cord.
4. There is a conglomeration of fibers that extends from the vestibular nucleus to the nucleus of the other cranial nerves, particularly the oculomotor nucleus. Connection of the cochlear root is by fibers that leave the anterior auditory nucleus and run transversely toward the midline, crossing the superior olivary nucleus to reach the olivary nucleus on the opposite side. This forms what is known as the *trapezoid body* (located in the inferior third of the pons). The fibers that depart from the lateral acoustic nucleus run into the *floor* of the fourth ventricle, forming the acoustic strias, integrating the most lateral part of the lateral lemniscus. They extend upward with the lemniscus to end in the inferior quadrigeminal tubercle, sending some fibers to the anterior quadrigeminal tubercle, and from here, follow an ascending course, entering the posterior limb of the internal capsule to end at the first temporal convolution.

5. Superior olivary nucleus. This is located at the dorsal surface of the outer third of the trapezoid body. Many fibers that cross the midline to enter the accessory auditory nucleus of the opposite side are derived from the superior olivary nucleus.

Besides the cranial nerve nucleus, we also have a conglomeration of cells forming the *nucleus pontis*. Among the longitudinal fibers are a tactile as well as a *proprioceptive tract*.

1. The ventral spinothalamic tract in the pons, becomes incorporated in the medial lemniscus.
2. The fasciculus gracilis and fasciculus cuneatus carries, in addition to tactile impulse, impulses from muscles, tendons, and joints. Subsequently, this fasciculus integrates the medial lemniscus previously described.
3. The dorsal spinocerebellar tract, from the dorsal nucleus of Clarke, has axons that extend upward and enter the pons. They are located in the center of the pons and run upward to the upper third of the pons, curving backward to enter the cerebellum via the superior cerebellar peduncle to end in the culmen, pyramid, and uvula.

Among the descending fibers is a pyramidal tract, already discussed, the tactospinal tract, and the rubrospinal tract.

Syndromes of the Pons   16

The pons is divided, from the anatomico-physiologic point of view, into thirds: superior, medial, and inferior. Each has two halves; a medial half and a lateral half. As a consequence of this, the following can be observed.

1. **Medial inferior pontine syndrome** (in general produced by occlusion of the paramedian branches of the basilar artery). It is characterized by:

    - Ipsilateral paralysis of the conjugate gaze to the side of the lesion due to destruction of the *para abducens* nucleus for conjugate lateral gaze.
    - Ipsilateral nystagmus due to destruction of *vestibular* nucleus and its connection.
    - Ispsilateral diplopia on lateral gaze due to destruction of the *abducent nerve*.
    - Contralateral paralysis of the face and extremity due to damage to the *corticobulbar* tract and the *corticospinal* tract.
    - Contralateral impaired tactile proprioception of the extremities and trunk due to damage to the *medial lemniscus*. This medial inferior pontine syndrome is known as *Millard-Gubler's syndrome*.

2. **Lateral inferior pontine syndrome** (due to occlusion of the anterior inferior cerebellar artery). This syndrome is characterized by:

    - Ipsilateral, horizontal, and vertical nystagmus due to damage to the vestibular nucleus.
    - Vertigo, nausea, and vomiting, also due to damage to the vestibular nucleus.
    - Ipsilateral facial paralysis due to damage to the facial nerve.
    - Ipsilateral paralysis of conjugate gaze due to damage to the para abducen center, for conjugate lateral gaze.

- Ipsilateral deafness, tinnitus due to damage to the auditory nerve and cochlear nucleus.
- Ipsilateral ataxia due to damage to the middle cerebellar peduncle and cerebellar hemisphere.
- Ipsilateral impaired facial sensation due to damage to the descending tract and nucleus of cranial nerve V.
- Contralateral impaired pain and temperature on half of the body due to damage to the spinothalamic tract.

3. **Medial mid-pontine syndrome** (due to occlusion of the paramedian branch of the basilar artery). It is characterized by:

   - Ipsilateral ataxia of limbs and gait due to damage to the middle cerebellar peduncle.
   - Contralateral paralysis of face, arm, and leg due to damage to the corticobulbar and corticospinal tract.

4. **Lateral mid-pontine syndrome** (due to damage to the short circumferential branches of the basilar artery). This syndrome is characterized by:

   - Ipsilateral limb ataxia produced by damage to the middle cerebellar peduncle.
   - Ipsilateral paralysis of muscles of mastication produced by damage to the motor nucleus and tract of cranial nerve V.
   - Ipsilateral facial sensation produced by impairment of the trigeminal (sensory) nucleus and tract.

5. **Medial superior pontine syndrome** (it is produced by occlusion of the paramedian branch of the basilar artery). This syndrome is characterized by:

   - Ipsilateral cerebellar ataxia due to impairment of the superior and middle cerebellar peduncle.
   - Ipsilateral internuclear ophthalmoplegia produced by damage to the medial longitudinal fasciculus.
   - Myoclonus of the soft palate, pharynx, and vocal cords due to damage to the central tegmental tract.
   - Contralateral paralysis of the face and extremities produced by damage to the corticobulbar and corticospinal tracts.
   - Contralateral impairment of touch, position, and vibratory sense due to a lesion in the medial lemniscus.

6. **Lateral superior pontine syndrome** (is due to occlusion of the superior cerebellar artery). This syndrome is characterized by:

   - Ipsilateral ataxia of limbs and gait, the patient will fall to the side of the lesion. It is produced by damage to the middle and superior cerebellar peduncles and superior surface of the cerebellum.

- Dizziness, nausea, and vomiting produced by damage to the vestibular nucleus.
- Nystagmus produced by damage to the vestibular nucleus.
- Ipsilateral Horner due to damage to the descending sympathetic fibers.
- Contralateral impaired sensation for pain and temperature in the face, limbs, and trunk due to damage to the spinothalamic tract.
- Contralateral impaired touch, vibration, and position sense due to damage to the medial lemniscus.

The Medulla Oblongata 17

The *medulla oblongata* extends from the inferior portion of the pons to a plane passing through 1 cm below the decussation of the pyramidal tract. From that point on, the *spinal cord* begins. The previously described plane corresponds to the inferior lip of the *foramen magnum*. It has four surfaces: one anterior, one posterior, and two lateral, left and right. The anterior aspect of the medulla oblongata, as does the pons, rests on the *basilar canal*, or *clivus*. The *posterior* aspect forms the lower part of the fourth ventricle and has an intimal relationship with the cerebellum. In the *anterior* aspect, in the midline, we find an *anterior fissure*, which below is in continuation with the anterior sulcus of the spinal cord. In the upper part, the anterior fissure ends in a blind pouch known as the *foramen cecum*. At the most inferior portion, the anterior fissure is interrupted by the fibers of the decussation of the pyramidal tract. The posterior and medial fissure is very well demarcated in the lower third of the medulla oblongata and continues above with the fourth ventricle.

These two fissures divide the medulla oblongata into two symmetrical halves. In the *anterior* aspect of the medulla oblongata, one can see the emergence of some of the cranial nerves of the medulla. The medulla could be divided into three columns:

1. The most medial column is located between the medial fissure and fiber of origin of the *hypoglossal nerve* forming the pyramids.
2. The lateral column corresponds to that portion of the medulla located between the fibers of the hypoglossal nerve and the fibers of the glossopharyngeal nerve, the vagus nerve, and the spinal accessory nerves.
3. In the lowest part of the medulla oblongata, the lateral column is single and receives the name of *lateral tract*.

In the upper part of the medulla oblongata one can identify the *olivary body*. The *posterior column* is that portion which is situated between cranial nerves IX, X, and XI, and the posterior and medial fissure. They are known with the name of *fasciculus cuneatus* and the *fasciculus gracilis*. These two fascicles join together forming a single structure known as restiform bodies, right and left. The pyramids are located at each side of the midline, separated from the preolivary sulcus where the root of cranial nerve XII emerges. It should be mentioned that the uncrossed pyramidal tract lies very close to the anterior and medial fissure. The lateral column of the medulla oblongata is in direct continuation with the lateral column of the spinal cord.

The *olivary body* is an oval-shaped mass situated at the lateral portion of the pyramid. Posteriorly, the olive is separated from the inferior cerebellar peduncle by the fibers cranial nerves IX, X, and XI. Inferior to the olive, one can find the *substantia gelatinosa of Rolando*, where the sensory fibers of cranial nerve V end. The *fasciculus gracilis* is a direct continuation of the posterior and medial column of the spinal cord and is separated laterally from the fasciculus cuneatus by a very superficial sulcus.

The *restiform bodies* occupy the upper part of the posterior surface of the medulla oblongata and are the direct continuation upward of the fasciculus cuneatus and gracilis. *Corpus restiformus* are crossed by transverse fibers known as *arcuate fibers*. When the two restiform bodies ascend upward and laterally, they form the lateral boundaries of the fourth ventricle. They subsequently curve backward and form the *inferior cerebellar peduncle* to enter the cerebellar hemisphere.

FOURTH VENTRICLE

At this particular time, it is appropriate to describe the fourth ventricle. It has a lozenge shape composed of two triangles joined at their bases. The *upper* triangle corresponds to the *posterior* aspect of the *pons* and the *lower* triangle corresponds to the *medulla oblongata*. The vertex of the superior triangle continues with the aqueduct of Sylvius. The inferior angle is known as *calamus scriptorius*, the lateral angles form the *lateral recesses*. The fourth ventricle has a floor and a roof. The roof of the fourth ventricle is integrated by the two superior cerebellar peduncles, the superior medullary velum, the inferior medullary velum, the inferior tela chorodea, the lingula, and the obex.

The superior cerebellar peduncle, which represents the main cerebellar outflow, emerges from the cerebellar hemisphere and runs in an upward and forward course forming the upper and lateral boundaries of the upper half of the fourth ventricle. They approach each other as they reach the quadrigeminal plate and extend underneath this anatomic structure and decussate, as previously described, to embrace the red nuclei.

The *superior medullary velum* is a lamina of white matter that joins the two superior cerebellar peduncles, forming an important part of the superior portion of the roof of the fourth venticle. The *inferior medullary velum* is another thin layer of white substance that extends from the nodulus to the ependyma of the floor of the fourth ventricle. It is continuous with the white matter of the cerebellum.

The *inferior tela choroidea* is a layer of the pia matter, which covers the lower part of the fourth ventricle just below the inferior medullary velum. In the upper portion, it is reflected under the surface of the cerebellum.

Inferiorly, it is in continuation with the restiform bodies and inferior portion of the medulla oblongata. Between the two layers of the tela choroidea, is a pair of choroid plexuses. These choroid plexuses have the shape of an inverted L with a vertical and a lateral portion. The lateral portion extends to the lateral recess of the fourth ventricle. The tela choroidea has an opening in the midline known as *foramen of Majendie*, and a lateral opening known as *foramen of Luschka* through which the fourth ventricle communicates with the subarachnoid space.

LINGULA

The *lingula* are very narrow bands of white matter that extend from the internal border of the corpus restiformis and the internal border, which is in continuity with the epithelial roof of the fourth ventricle.

FLOOR OF THE FOURTH VENTRICLE

The floor of the fourth ventricle has a *rhomboid* shape. It is divided into a left and right half by a *medial sulcus*. Traversing the fourth ventricle we find the *stria medullaris* or *stria acoustica*. These striae divide the floor of the fourth ventricle into two triangles, superior and inferior. The inferior triangle represents the lower half of the floor of the fourth ventricle. The superior triangle represents the upper half of the floor of the fourth ventricle. Above the stria acoustica there is a slight depression of the fourth ventricle known as the *superior fovea*. The superior fovea has a triangular shape and is limited superiorly and medially by the stria of the harmony; inferiorly, by the stria acoustica; and laterally, by the superior cerebellar peduncle. This portion is located in the pontine triangle of the fourth ventricle. Just below the previously described depression in the inferior triangle of the fourth ventricle, we find the *trigonum acousticus*.

In the *inferior triangle*, in the midline, there are two eminences known as trigonum of the hypoglossal nerve. These eminences have two portions, one more medial, known as *area medialis*, and one lateral, known as *area plumiformis*. The trigone of the hypoglossus is also known as *white internal ala*. Lateral to the hypoglossal trigone, there is a depressed triangular area of ashy color known as the *area cinerea*—which becomes somewhat elevated at the most inferior part. This area is also known as the *trigonum of the vagus nerve* and corresponds to the nucleus of the vagus and glossopharyngeal nerve, also known as the *gray ala*. Lateral to the gray ala, there is a slight prominence of the ependyma, which also has a triangular shape and is known as the *area postrema*.

The *superior triangle* has, immediately above the trigone of the hypoglossum, a longitudinal eminence known as *teres eminence*. Below this, one can find the nucleus of cranial nerve VI, as well as the fibers of cranial nerve VII (as they wrap around the nucleus of cranial nerve VI). The *teres eminence* continues upward with a longitudinal-shaped bundle known as *foniculus incertus*.

Above each superior fovea and lateral to the teres eminence, there is a grayish-bluish area known as *locus ceruleus*. This color is due to ferruginous pigment; and the location is one of the roots of cranial nerve V termination.

Internal Structures of the Medulla Oblongata

As two of the cranial nerves exit the medulla oblongata, the hypoglossal nerve (medially) and the vagus nerve (posterolaterally), this divides the internal structure of the medulla oblongata into three portions (one anterior, one intermedial, and one posterolateral).

The *anterior* portion contains primarily the pyramidal tract, the medial lemniscus, the accessory olivary nucleus, the medial tectospinal tract, and the dorsal longitudinal fasciculus.

The *lateral* area, situated between the hypoglossal nerve (medially) and the fibers of the glossopharyngeal and vagus nerve (laterally), contains the olive, also known as *olivary nucleus*. The inferior olivary complex is divided into the inferior olivary nucleus itself, an accessory olivary nucleus, and a dorsal accessory olivary nucleus. The main inferior olivary nucleus has a U-shaped nucleus, which has a convoluted serpentine structure with the opening of the U towards the midline—this part is known as the *hilus*. This nucleus has two layers known as *lamellas*, a dorsal and a ventral one. The cell types of the inferior olivary complex are small and also spherical. Many of its fibers extend to different areas of the spinal cord. Most of the cells project into the ovilary cerebellar tract. It seems that many authors comment that the medial part of the olivary nucleus is connected with the inferior vermis of the cerebellum. The inferior olivary nucleus is connected with the surrounding reticular nuclei of the medulla oblongata. A very important connection of the inferior olivary complex is the *central tegmental bundle*—which is connected with the thalamus. A group of fibers extends from the red nucleus to the inferior olivary nucleus. Lateral and posterior to the inferior olivary nucleus is the *vagal nucleus*, or *nucleus ambiguous*. It is formed by a nuclear column that extends from the medulla oblongata to the pons. From there, we have dorsal afferent fibers that contain preganglionic fibers that supply various viscera, particularly the heart and the respiratory system. The parts of the nucleus that provide innervation to the abdominal viscera are situated in the most caudal portion of the nucleus ambiguous. Parasympathetic fibers from the viscera form a conglomeration of afferent components, which when they enter the nerves in the medulla oblongata, accumulate in a bundle known as fasciculus solitarius, which is located posteriorly and laterally to the dorsal portion of the nucleus ambiguous. Posterior and medial to the *fasciculus solitarius* is the *posterior salivatory nucleus* to the *nucleus ambiguous*.

The *posterior* area comprises that portion situated between the fibers of cranial nerves IX, X, and XI, and the posterior medial fissure. In that portion, we find posterior and medial to an anterior and lateral direction, the accessory nucleus of the vagus, the fasciculus solitarius, the nucleus gracilis, the nucleus cuneatus, the corpus restiformis, and the nucleus gelantinosus of Rolando. In the *lateral* angle of the fourth ventricle, sharing functions with the pontine nucleus, is the *dorsal cochlear nucleus* and the *ventral cochlear nucleus*, which are spread out through the medial lemniscus. It is a conglomeration of fibers that extends, after it enters the lateral lemniscus, to the one on the opposite side to form the trapezoid body. The lateral lemniscus extends in an upward course to the inferior colliculus. The funiculus Rolando, which is the sensory nucleus of cranial nerve V, is separated from the surface of the medulla oblongata by a band of fibers known as the ascending root of cranial nerve V, as well as by the arcuate fibers. Above the nucleus gelatinosis is found the nucleus medial sensitive of the trigeminal nerve. From the nucleus gelatinosis, as well as from the medial nucleus, there is a conglomeration of fibers that runs toward the midline and then decussates in the *raphe* of the pons with the opposite side to follow an upward direction. Then, the fibers extend towards the pons, joining the medial lemniscus and then end in the optic thalamus. During this, the ascending fibers have *collaterals* that extend in the *bulbopontine motor nucleus*.

RETICULAR FORMATION

The *reticular formation* of the medulla oblongata is located behind the pyramidal tract and the olivary bodies, and extends laterally to the restiform bodies. It has, as the name implies, a *reticular* appearance due to the intersection of bundles of *longitudinal* as well as *transverse* fibers. In the most *anterior* portion, there is almost an *absence* of nerve cells. While in the *lateral* area, there are *multiple* nuclei of nerve cells known as the *gray reticular formation*. One important group of these nuclei is located in the dorsal aspect of the hilum of the inferior olive and is known as *inferior central nucleus*. The transverse fibers of the reticular formation are known as the *internal arciform fibers*.

In summary, the nucleus of the *hypoglossal* nerve, the *auditory* nuclei (which have a dorsal or internal auditory nucleus located external to the glossopharyngeal nucleus), *originate in* the medulla oblongata. There is another nucleus known as the *ventral auditory nucleus* and another conglomeration of nerve cells outside of the restiform body known as the *lateral acoustic tubercle*. The nucleus of the glossopharyngeal and pneumogastric nerve are two, a principal and an accessory. The main nucleus of these two nerves is located just below the ependyma of the fourth ventricle at the gray ala. The *accessory nuclei* are located in the reticular formation and are known as *nucleus ambiguous*. The nucleus of the accessory nerve is located at the lower part of the medulla oblongata and beneath the ependyma of the fourth ventricle, laterally to the nucleus of cranial nerve XII.

Understanding the anatomy of the medulla oblongata, it will then be easier to understand the medullary syndromes. The medullary syndromes can be divided into three types as follows.

Medial medullary syndrome, also known as anterior preolivary syndrome.

It is produced by occlusion of the short medial branch of the vertebral artery. It is characterized by:

- Ipsilateral paralysis, with atrophy of the tongue due to damage to cranial nerve XII.
- Contralateral paralysis of the arm and leg, with facial sparing, due to damage to the corticospinal tract.

- Impaired tactile and proprioception over half of the body due to damage to the medial lemniscus.

Lateral medullary syndrome, also known as Wallenberg syndrome. This syndrome is produced by occlusion of the posterior inferior cerebellar artery. It is characterized by:

- Ipsilateral loss of pain in half of the face due to damage to the trigeminal tract nucleus.
- Ipsilateral nystagmus and diplopia due to damage to the vestibular nucleus and connection.
- Vertigo, nausea, and vomiting, due to damage to the vestibular nucleus and connection.
- Dysphagia and hoarseness due to vascular injury to the fibers of cranial nerves IX and X and nucleus ambiguous.
- Ipsilateral paralysis of vocal cord due to damage to the fibers and nucleus of cranial nerve X.
- Ipsilateral diminished gag reflex, due to damage to cranial nerves IX and X. There is ipsilateral paresis of the soft palate, pharynx, and larynx. The patient presents with a raspy and twining voice.
- Ipsilateral myosis and ptosis (Horner's syndrome) produced by vascular insult to the descending sympathetic tract.
- Ipsilateral ataxia due to vascular injury to the cerebellum and spinocerebellar pathway.
- Contralateral loss of pain and temperature due to vascular injury to the spinothalamic tract.

Posterior (or retro-olivary) syndrome.
Posterior syndrome is a rather uncommon syndrome that involves the nuclei cranial nerves IX, X, XI, and XII. It should be remembered that the pyramidal tract is never affected. Due to the variety of combination of the different cranial nerve involvement, for the sake of completeness, I shall mention the so-called *Schmid syndrome*, which involves cranial nerves IX, X, and XI. *Tapia syndrome* involves cranial nerves IX, X, and XII. Finally, *Jackson's syndrome* involves cranial nerves IX, X, XI, and XII, with paralysis. Posterior retro-olivary nucleus syndrome, as well as the others mentioned, are provoked by vascular occlusion of the short circumferential arteries.

Another rather uncommon syndrome is known as *Babinski-Nagotte*, which is indeed damage to *half* of the *medulla oblongata*. It can be produced by incomplete *occlusion* of the *vertebral artery* and results in contralateral hemiplegia, alternating hemianesthesia, ipsilateral ataxia, and ipsilateral Horner's syndrome. It should be stressed that the cranial nerves are NOT affected in this syndrome.

The Cerebellum 18

The cerebellum is located in the posterior fossa, which has as its upper limits the *tentorium*, and it is concerned with *coordination* of motor response automatism. The cerebellum is subdivided into a central portion known as the *vermis*, with two *cerebellar hemispheres* on each side of the vermis. It consists of white and gray matter. The external surface is traversed by numerous sulci that separate it in laminous-like structures. Under each cerebellar hemisphere, there is a small area known as the *flocculus*. The flocculus adjoins the vermis and forms the *paleocerebellum*. The cerebellar hemispheres are known as the *neocerebellum*.

The *vermis* is divided into two portions, the superior and inferior vermis. The vermis has different portions, beginning from the most superior and anterior, there is the *lingula*. Behind this is the *precentral fissure*. Next, we find the *central lobe*—behind which is the postcentral fissure. Then, behind the postcentral fissure is the *culmen*. Posterior to the culmen is the *preclival fissure*.

Behind the **pre**clival fissure is the *clivus*. Immediately behind the clivus is the ***postclival fissure***. Posterior to the postclival fissure is the *folium*, behind which is located the great horizontal fissure. Immediately behind this fissure is the *tuber*. Behind the tuber is the postpyramidal fissure. In front of the **post**pyramidal fissure is the *pyramis*; and in front of the pyramis is the **pre**pyramidal fissure. The *uvula* is immediately in front of the prepyramidal fissure, but behind the postnodular fissure. Finally, in front of the postnodular fissure is the nodule. In synthesis, the vermis has from superior and anterior, to inferior and anterior: the lingula, central lobe, culmen, declivus, folium, tuber, pyramis, ulvula, and nodule.

In the cerebellar hemisphere, the vermis continues as follows: the central lobe continues laterally with the *ala centralis*, the *culmen* with the anterior semilunar, the *clivus* with the posterior semilunar (also known as the *crescentic lobe*), the *folium* by the posterior and superior lobe, and the tuber by the posterior and inferior lobe. These last two structures are separated by the great horizontal fissure. The pyramis continues laterally with the *digastric lobe*, the uvula with the tonsil of the cerebellum, and the nodulus with the flocculus. In the posterior surface of the cerebellum one can identify the postcentral sulcus, the preclival fissure, the postclival fissure, the great horizonal fissure, the post- and prepyramidal fissures, and the postnodular fissure can be identified. In front of the viventer lobe, and behind the posterior inferior lobe (in connection with the tuber), is the gracilis lobe. In the inferior vermis, the nodule and the flocculus project into the roof of the fourth ventricle, and it can be seen clearly after the cerebellum is separated from the pons. On each side of the nodule, there is a thin membrane of white substance known as the *inferior medullary velum*. Laterally, it continues with the white matter of the cerebellum.

The *uvula* and the *tonsils* occupy a great portion of the inferior vermis. The tonsils are located in a deep fossa between the uvula and the viventer lobe. It should be mentioned that the inferior vermis, including the pyramid and the nodulus are *absent* in the Dandy-Walker syndrome. Also absent in this syndrome is the *posterior inferior cerebellar artery* (PICA).

As stated previously, the cerebellum consists of gray and white matter. In a sagittal section, there is a mass of *gray* matter known as *corpus dentatum* (dentate nucleus or cerebellar olives). The *white* matter of the cerebellum includes the *peduncular fibers*, which are in direct continuity with the cerebellar peduncle and the fibers proper of the cerebellum.

From the anterior part of each cerebellar hemisphere, originate three large columns of white matter known as: superior, medial, and inferior *cerebellar peduncles*. These connect the cerebellum with the cerebral peduncles, pons, and medulla oblongata.

The *superior* cerebellar peduncles, mentioned before in the description of the fourth ventricle, form the superior and lateral boundaries of the fourth ventricle. They extend forward and backward toward the cerebral peduncle beneath alas of the central lobe, extending to the interior of the cerebellar olive and dentate nucleus. They decussate with the fibers of the opposite side below the quadrigeminal plate and wrap around the red nucleus, forming the *white nucleus*, to reach the thalamus at the ventrolateral nucleus. In the superior cerebellar peduncle, there are also fibers of the spinal cord that curve backward to enter the cerebellar hemisphere. In between the two superior cerebellar peduncles, there is a transparent lamina of white matter known as *superior medullary velum*, forming with the former the upper roof of the fourth ventricle.

The *middle* cerebellar peduncles are the largest of the three pears. They consist of large transverse fibers that extend from one cerebellar hemisphere to the other forming most of the transverse fibers of the pons. They enter the cerebellum, extending to the upper and lower part of the hemisphere to the cerebellar cortex.

The *inferior* cerebellar peduncles connect the cerebellum with the medulla oblongata. From the level of the clava, they follow an upward and outward course forming part of the lateral boundaries of the floor of the fourth ventricle, and enter the cerebellum below the middle cerebellar peduncle. Then, they follow an upward course—ending in the cortex of the cerebellar hemisphere and superior vermis. The inferior cerebellar peduncle also has fibers that connect the spinal cord with the cerebellum. These are the direct spinocerebellar tract; fibers from the fasciculus, nucleus cuneatus, and gracilis; fibers from the olivary nucleus of the medulla oblongata; and fibers of cranial nerves V, VIII, IX, and X.

As mentioned before, the so-called *fiber propria of the cerebellum* are the *commissural* fibers and the *arcuate* fibers, which connect the adjacent lamina with the one on each side. The cortex of the cerebellum has a foliated appearance and is arranged in three layers:

1. A molecular layer that has cerebellar cells as well as fibers. These fibers are derived from the axons and then dendrites of the Purkinje cells.
2. Fibers from cells of the granular layers.
3. Fibers from the central white substance of the cerebellum.

The cells of the *molecular* layer are rather small. They have long cylindrical fibers that run a long horizontal course from which there are a great deal of collateral branches that change to a vertical course toward the Purkinje cells, creating around those cells a basket-like collection—for which reason they are named *basket cells.*

The corpuscles of Purkinje, or simply *Purkinje cells*, have a pear shape and they are located between the molecular and granular layers. From the vertex of the pear a lot of axons arise, they pass through the external granular layer, and continue in the medullary substance beneath. The internal granular layer is characterized by numerous, small granular cells with many nerve fibrils. Most of their axons extend into the molecular layers and bifurcate in right angles to run a horizontal course. In addition, in the most outer part of the granular layer, are larger cells known as *Golgi cells.* Their axons enter the granular layer, while the dendrite extends to the molecular layer.

In the *gray* matter of the cerebellar cortex a great deal of fibers, which originate in cells that, it is believed, proceed from the spinal cord, can be seen. These branches are subdivided in numerous branches. These fibers are known as the "moss" fibers that form an arboresence around the granular layer. Some of these branches go around the dendrite of the Purkinje cells and are known as *clinging fibers.*

In the cerebellar *white* matter, one can also see four nuclei of gray matter, one is rather large and there is one for each hemisphere (known as *nucleus dentatum*). The other three are known as: *nucleus emboliformis, nucleus globosus,* and *nucleus fastigii.*

The *nucleus dentatum* is located on each side of the midline, in the center of the white matter of the cerebellum. It consists of a conglomeration of folded laminas with an open hilum located at the most internal portion, from which most of the fibers of the superior cerebellar peduncle emerge.

The *nucleus emboliformis* is a mass located at the inner portion of the dentate nucleus and is partially covering the hilum.

The *nucleus globosus* is located in the inner side of the nucleus emboliformis.

The *nucleus fastigii* is located in the midline at the anterior end of the superior vermis, just below the lingula, over the roof of the fourth ventricle—which is why they are known as *nuclei of the roof.*

The anterior lobe of the cerebellum receives the dorsal as well as the ventral spinocerebellar tract. This dorsal spinocerebellar tract originates from the *dorsal column of Clarke* and enters the cerebellum through the inferior cerebellar peduncle. It distributes in the culmen, clivus, pyramids, and uvula. The inferior olivary nucleus, from the medulla oblongata as well as the accessory olivary nucleus, has fibers that are direct, as well as others that are decussated, and enter the cerebellum via the inferior cerebellar peduncle. These fibers project to the flocculus as well as the posterior lobe of the cerebellum. The vestubular nucleus and the vestibular system project through the inferior cerebellar peduncle and also into the flocculonodular system and uvula. They are an important connection of the trigeminal nuclei with the cerebellum, and also through the inferior cerebellar peduncle to distribute into the anterior and posterior cerebellar lobe.

A very important group of fibers that connect the nucleus fastigii, following the superior cerebellar peduncle and curving back and downward to enter the brain stem, pons, and spinal cord, is known as the bundle of hook or Russell.

The premotor cortex sends a large number of afferent fibers that follow the internal capsule entering the cerebral peduncle and decussate to the opposite pontine nuclei. From there, they extend to the cerebellar cortex with the name of pontocerebellar fibers, in essence forming the corticopontocerebellar tract; from the cerebellar hemisphere. It establishes with the ipsilateral nucleus of the cerebellum, and from there, it leaves the cerebellum through the superior cerebellar peduncle, crosses to the opposite side to involve the red nucleus, to end in the ventrolateral nucleus of the thalamus. From the ventrolateral nucleus of the thalamus, it extends to the parietal and premotor cortex, through the posterior limb of the internal capsule.

Fibers from the superior and the inferior colliculus extend to the cerebellum and to the vermis forming the tectocerebellar tract via the superior cerebellar peduncle.

The *flocculonodular system* is the area of the cerebellum where the vestibular fibers terminate. The flocculus sends impulses through its connection to the superior and lateral vestibular nucleus as well as the oculomotor nucleus through the superior cerebellar peduncle; some of the fibers join the medial longitudinal fasciculus.

THE PATHOPHYSIOLOGY OF THE CEREBELLUM

Many authors believed that the cerebellum was primarily the organ of the *equilibrium*, although it does participate in the equilibrium, it has great participation in the so-called *check mechanism*. The cerebellum is the great organ of the *proprioceptive system*. Several clinical manifestations can be observed in lesion of the cerebellar hemisphere. One is known as *asthenia*, manifested by loss of strength in the muscles located at the ipsilateral lesion of the cerebellar hemisphere. Another is *ipsilateral atonia*, manifested by diminution of the muscular tone. In a lesion of the cerebellar hemisphere affecting a great part of the cerebellar hemisphere, terminal *tremor* is observed in the test "finger-to-nose" and "heel-to-knee." Every cerebellar hemisphere acts predominantly over the *ipsilateral* muscles.

The *vermis* affects the movement of the facial muscles, eyes, neck, and trunk. The *flocullonodular* system exerts control over the movement of the tongue, lips, and throat. It must be clearly stated that the *superior* and *posterior* portion of the cerebellar hemisphere controls the *upper* extremities; while the *anterior* and *inferior* controls the *lower* extremities.

A lesion in the most *anterior* part of the *superior vermis* produces a tendency to fall *backwards*. On the other hand, if the lesion is located at the most *inferior* and *posterior* portion of the *vermis*, the patient tends to fall *forward*.

It could be stated that the eumetry of the muscles is produced by the cerebellar hemisphere. In order to have coordination of the different muscles when an action is performed (synergism), the integrity of the cerebellar hemisphere is necessary. It must be remembered that the cerebellum has to receive, first of all, the information from the cerebral cortex via the *corticopontocerebellar tract* for this synergism to take place. Interacting with the cerebellum are the impulses coming from the vestibular nucleus as well as with the rubrospinal and vestibulospinal tract. It is a rather complex phenomenon, but it can be established that *damage* to the *cerebellum* produces impairment of muscle tone.

A lesion that involves the *flocculonodular system* is by far the *most common cerebellar disorder in children*, producing ataxia and a broad gait. This disorder is primarily produced by a tumor of the vermis that compresses the floor of the fourth ventricle. Usually, it is accompanied by nausea, vomiting, and headaches (due to increased intracranial pressure because the mass produces blockage of the circulation of the cerebrospinal fluid, creating *hydrocephalus*). In my experience, in young children 5 years of age or younger, a common cause of *gastric outlet obstruction* is produced by a tumor of the *vermis*, which compresses the floor of the fourth ventricle.

A lesion in the floor of the fourth ventricle originating from the cerebellum can create *positional paroxysmal nystagmus*. Positional paroxysmal nystagmus should alert the examiner, who should think about a lesion that affects the *fourth ventricle*. At times, a lesion of the vermis can also produce nystagmus in all directions, known as *perverted nystagmus*.

When there is a lesion affecting the *anterior* portion of the cerebellar hemisphere, the patient presents with what is known as *cerebellar fits*, characterized by rigidity and arching of the back.

A tumor or lesion that affects the *posterior* cerebellar hemisphere produces a tendency to *fall towards the same side* of the lesion, and there is a loss of the check mechanism. Therefore, when the strength of the upper extremity is examined, one should be careful because when the examiner lets the arm go, the patient could hit his or her own face. Lack of coordination is also noted in the "heel-to-knee" and "finger-to-nose" exam. The patient has a tendency to exhibit bradykinesia (slow movement) and/or adiadochokinesis. In the latter, the examiner requests the patient to rapidly rotate the hands. This is observed *ipsilateral* to the lesion. The speech may also be affected and usually is slurred and explosive.

Tremors that occur in the extremities *ipsilateral* to the lesion are usually *intentional* (whereas the tremor of a *parkinsonian* patient is *a tremor at rest*).

When a lesion occurs in the *dentate* nucleus, the patient has *intentional* tremors *contralateral* to the affected area. It should be remembered that the superior cerebellar peduncle decussates at the level of the cerebral peduncle.

In syndromes of the cerebellum, the patient may have subjective as well as objective symptoms and signs. Among the subjective findings are vertigo or dizziness. It usually occurs when the patient is standing, although it might also occur when lying in bed. Headaches and vomiting would be present. The objective signs are: *tremors*, which are usually *intentional*; *oscillation* when standing; and intent to *increase* the *distance* between their feet, increasing the sustentation base. However, if the patient closes his or her eyes, he or she does NOT fall. So, it could be stated that in pure cases of cerebellar syndrome, the *Rhomberg sign* is *never* present. As stated previously, the patient can have a tendency to fall forward or backward, there is marked impairment of the gait and when walking, he or she has a tendency to exaggerate the flexion of the extremities. Therefore, the patient raises the legs higher than necessary and on many occasions, the patient appears as if he or she is drunk.

Another disorder that is observed is difficulties in writing due to tremor and dysmetria. Usually the letters are large (macrographia). It should be remembered that the parkinsonian patient, in contraposition, presents with small letters (micrographia).

CEREBELLAR SYNDROME

There are many causes or etiological factors for *cerebellar syndrome* as follows:

1. Tumor of the cerebellum (which can occur in the vermis or the cerebellar hemisphere). The most common tumor of the cerebellar hemisphere in the adult, is *metastasis*. The most common *primary* tumor of the cerebellar hemisphere in the adult is a *hemangioblastoma*, followed by *astrocytoma*. Hemangioblastomas are associated with diseases of the neuroectodermal system (like Von Hippel-Lindeau) and ataxia telangectasia. In *children*, the most common tumor of the cerebellar *hemisphere* is a *cystic astrocytoma*, and the most common tumors of the cerebellar *vermis* are *medulloblastoma*, ependymoma, terratoma and sarcoma.
2. Cerebrovascular accident.
3. Cerebellar trauma.
4. Cerebellar abscess.
5. Acute cerebellar ataxia produced by viral infection.
6. Primary degenerative diseases of the cerebellum such as olivopontocerebellar degeneration.
7. It should be remembered that due to the connection between the frontal lobe and the cerebellum via corticoponticocerebellar tract, tumor of the frontal lobe can also produce cerebellar ataxia known as Brun's ataxia.
8. Multiple sclerosis.

9. Friedreich's ataxia. This is the most common of the spinocerebellar ataxias. This ataxia is characterized by degeneration of the spinocerebellar tract and corticospinal tract, as well as the posterior column of the spinal cord. In the cerebellum, although the changes are variable, atrophy of the Purkinje cells and dentate nucleus can also be found.
10. A syndrome, cerebellosum, can also be observed in a lesion that involves the pons and medulla oblongata
11. Dandy-Walker syndrome is a congenital disorder due to occlusion of the foramen of Luschka and Magendie. This syndrome is accompanied by increased intracranial pressure and hydrocephalus. Cranial nerves IX, X, XI, and XII can be affected. Due to the cystic dilatation of the fourth ventricle, there is marked compression of the cerebellum and brain stem.
12. Arnold-Chiari malformation, particularly types II and III.

If a patient (beside cerebellar manifestations) also has involvement of the pyramidal tract, very often it is due to *multiple sclerosis*.

The Cranial Nerves 19

There are a total of *twelve* cranial nerves. They will be described one by one. In the description of the cerebral peduncle, pons, and medulla oblongata, the different nuclei of origin of the cranial nerves have already been described (see Figures 19-1 and 19-2).

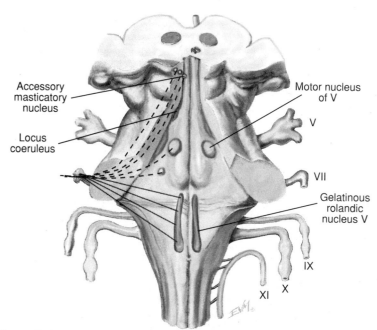

Figure 19-1 The brain stem and the cranial nerves.

98 NEUROSCIENCE

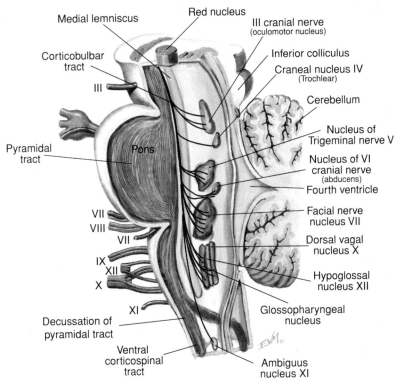

Figure 19-2 The cranial nerves and nuclei at the brain stem level.

OLFACTORY NERVE

Cranial nerve I, also known as the *olfactory nerve*, is the one in charge of perception and of smell in general. It is integrated by a conglomeration of small nerve fibers that emerge from the olfactory bulb, which is located at the cribiform plate at the ethmoid bone. From the olfactory bulb, they have a descending course, traveling through the different orifices of the cribiform plate and extend to the superior portion of the nasal cavity, in two group of fibers, internal and external. These fibers distribute themselves in the superior and medial turbinates at the olfactory mucosa. Each nerve is surrounded by the extension of the dura mater. Each of the previously described nerves ends in the olfactory cells of Schultze.

The olfactory pathway is integrated, first by peripheral neurons, the cells of Schultze. The axons of these constitute the olfactory nerve that ends in the olfactory bulb. A second neuron is located at the olfactory bulb where synopsis takes place with the axon of the first peripheral neuron. At this second neuron, the olfactory tract begins. This olfactory tract is divided into a lateral and a medial branch. The lateral one extends to the amygdaloid nucleus in the hypocampus and the cortex of the hippocampus. The medial root follows the corpus callosum, genu, body, and splenium; at the level of the splenium it becomes thickened and becomes the fasciola cinerea, which will convert in fascia dentata to have an ascending course over the uncus of the medial aspect of the hippocampus, and receives the name of *tract of Giacomini*.

Alteration of the sense of smell can be grouped into several categories, including anosmia and hyposmia. *Unilateral anosmia* usually corresponds to a tumor at the level of the olfactory groove (like the olfactory groove *meningioma*), tumor of the frontal lobe, and trauma.

OPTIC NERVE

In reality, the optic nerve should be considered like a prolongation of the telencephalic vesicle, with a receptory portion that is the retina.

The apparent origin of the optic nerve occurs in the anterior and lateral portions of the optic chiasm (see Figure 19-3). From this area, it extends forward and lateral to reach the optic foramen to enter the orbital cavity. Along its course, the optic nerve has an intimate relation with the diaphragm sella, the carotid artery, and the ophthalmic artery. Once they enter the orbit, there

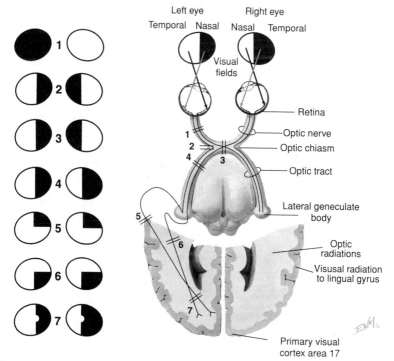

Figure 19-3 A pathway showing visual field defects.

is an intimate relationship with the muscle of the ocular globe. The ophthalmic ganglion and the ciliary nerve keep close contact to the optic nerve.

THE OPTIC PATHWAY

It has a peripheral neuron and two central neurons.

The peripheral neuron is represented by the bipolar cells, which occupy the medial portion of the retina and from which there are two axons. The *peripheral* axon is the one that perceives light impulses that run toward the rods and cones, and a *central* prolongation that runs toward the ganglion cells of the retina. From the ganglion cells of the retina originate the fibers that integrate the optic nerve (which reach the optic chiasm).

The *optic chiasm* has a rectangular shape and is a rather thin layer of white substance that has two surfaces, one superior and one inferior. The anterolateral angle of the optic chiasm receives the optic nerve. From the posterior and lateral angle of the optic chiasm two optic tracts originate and run in the lateral surface of the cerebral peduncle reaching the thalamus (where they are subdivided into internal and external bundles).

The *external bundle* ends at the *lateral* genicular body and in the pulvinar, and partially in the anterior and superior quadrigeminal tubercle. The *internal* subdivision disappears in the internal geniculate body and the posterior quadrigeminal tubercle. Not all the fibers of the optic tract end in the lateral and internal geniculate body, some of them run directly and reach the cerebral cortex. All of these fibers (the direct one as well as the one that stops at the geniculate bodies) run posteriorly toward the posterior portion of the internal capsule. It integrates what is known as the *optic radiation* of Gratiolet that ends at the cortex of the occipital lobe, at the area located around the calcarine fissure.

Not all the fibers that originate at the ganglionary cells of the retina follow a direct course in the optic chiasm. Instead, they suffer a partial decussation in that structure. As a result, the nasal part of the retina decussates completely at the center of the optic chiasm to reach the optic tract of the opposite side, whereas the temporal half of the retina runs without decussation in the optic chiasm to reach the optic tract of the same side.

Besides this group of fibers that originate in the nasal and temporal half of the retina, we find the *macular tract* which originates in the *macular lutea*. These fibers are partially decussated in the optic chiasm.

The clinical exploration of the optic nerve is best done by an ophthalmologist; however, it can be stated that the exploration of the optic nerve should involve visual acuity, color perception, visual fields, and examination of the eye grounds. When, by confrontation of the visual fields, one finds a disorder that consists of an inability to see in either the nasal or temporal fields, the patient presents with what is known as hemianopsia.

If the hemianopsia corresponds to homologous sides, we call this type *homonomous hemianopsia*. The *homonomous hemianopsias* are seen in the retrochiasmatic lesion involving the optic tract, the pulvinar, or the occipital lobe.

If, on the contrary, the defect corresponds to distinct different sides, one left half and one right half, then we call this *heteronymous hemianopsia*.

Bitemporal hemianopsias are due to lesions at the level of the optic chiasm, in the mid-portion where the decussation of the nasal retinal fibers occurs. This is described as *binasal visual field defect*. This visual field defect is commonly observed with a tumor of the pituitary gland. Binasal hemianopsia is a very rare occurrence and can only be found with a lesion that affects the most *lateral* side (symmetrically, right as well as the left) in the *optic chiasm*.

If the blind area is limited to a fourth of the visual field, it is called *quadrantianopsia*. If the blind area is above or below the horizontal plane of the eyes, it is called *horizontal* or *altitudinal hemianopsia*. It is observed in a lesion of the calcarine fissure or in uniform compression of the superior or inferior portion of the optic chiasm.

To examine the eye grounds it is necessary to use the ophthalmoscope. Many changes can be seen by means of this examination. We can observe *atrophy* of the papilla of the optic nerve. If the atrophy shows absolutely *white* characteristics with ill-delimited border of the papilla, one should think of *postneuritic atrophy*. In these cases, there is a disproportion between the size of the arteries and the veins. The arteries appear very slender and tortuous, whereas the veins appear markedly dilated and tortuous.

Inflammation of the optic nerve, known as neuritis optica, can be observed in many conditions such as syphilis, multiple sclerosis, and Devic's disease. Papillary edema is another anomaly that is oberved in the examination of the eye grounds. It is most commonly seen as a manifestation of increased intracranial pressure, as in hydrocephalus, malignant arterial hypertension, and brain tumors.

III, IV AND VI CRANIAL NERVES

These are the nerves that are in charge of the innervation of the extrinsic muscles of the *eyes* (see Figure 19-4).

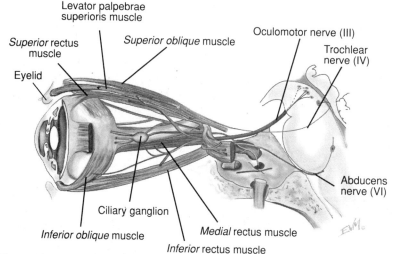

Figure 19-4 The cranial nerves to the eye muscles.

OCULOMOTOR NERVE

Cranial nerve III, also known as the *oculomotor* nerve, has an apparent origin at the *interpeduncular space*. As described in the section regarding the cerebral peduncle, the nucleus of origin of the cranial nerve is located at the *periaqueductal gray matter*. It runs forward and lateral, traversing the medial longitudinal fasciculus (MLF), the red nucleus, the white nucleus, and the internal border of the substantia nigra. It exits in between the posterior cerebral and the superior cerebellar artery, and follows a forward course to reach the posterior clinoid, entering the lateral wall of the cavernous sinus (together with the cranial nerve IV and the ophthalmic branch of the trigeminal nerve). Finally, it enters the orbit through the superior orbital fissure and divides into two branches, a superior and an inferior branch.

The *superior* branch gives the nerve supply to the superior rectus and elevator palpebral. The *inferior* branch supplies the inferior rectus, the inferior oblique, and the medial rectus. It also gives origin to fibers that traverse the cililary ganglion forming the short ciliary nerve, which innervates the pupillary sphincter and the cililary muscle. These two muscles have as a primary function, accommodation. The pupil is an orifice that is located in the central part of the iris and it has multiple connections with the cranial nerve and the central nervous system. It has a contractile system of muscles that allows a reduction or increase in diameter. The contractile system has two systems of fibers; one of them is arranged in a circular manner forming the pupillary sphincter, which is responsible for the contraction of the pupil. The other fibers have a radiating distribution forming the structure that dilates the pupil.

Although the pupil is normally evaluated by an opthhalmologist, during the patient examination, one should observe the shape or form, the size, and the location of the pupils.

The size is subject to the influence of light stimuli, but it has a normal size that fluctuates between 2 to 4 mm. The *size* of the pupil is regulated by cranial nerve *III* on one side, and by the *sympathetic* nerve on the other side. Cranial nerve *III* is responsible for the *contraction* of the pupillary sphincter, whereas the *sympathetic* nerve *dilates* the pupil. The pupil can either be dilated unilaterally and/or bilaterally. If *dilated*, we speak of *midriasis*, and if *constricted*, we speak of *myosis*. The fibers of the oculomotor nerve that innervate the circular sphincter of the pupil, as mentioned before, traverse the ciliary ganglion or ophthalmic ganglion forming the short ciliary nerve. These represent the parasympathetic system of the pupil and have their central origin in the nucleus of Edinger-Westphal, around the aqueduct of Sylvius.

The *sympathetic* fibers that supply the pupil originate in the spinal cord, at the intermedial lateral column of the cervicothoracic cord, known as the *ciliospinal* center. Fibers from this center, through the communicating rami of the cervical roots, C7-8, D1, D2, and D3, enter the cervical sympathetic system. And from there, they enter the inferior, medial and superior sympathetic ganglion. Afterwards, they form a conglomeration of fibers around the carotid arteries and reach the gasserian ganglion. This is where the ophthalmic branch of cranial nerve *V* and the nasal branch of the *ophthalmic* nerve give origin to the *long ciliary nerve* (that reaches the *pupillary sphincter*).

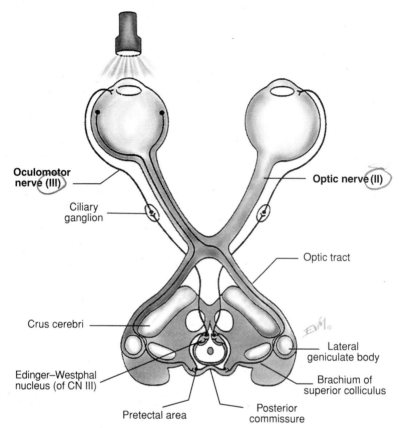

Figure 19-5 The pupillary light reflex pathway.

The reaction of the pupil to the light is known as the *photomotor reflex* (see Figure 19-5). Light stimulation of the rods and cones of the retina produce impulses that follow the optic nerve to the chiasm. Here, the pupillary fibers partially decussate. They subsequently follow the optic tract, the *lateral* geniculate body, and the anterior quadrigeminal tubercles. The center of the photomotor reflex is located at the level of these tubercles. From this center, there are pupillary fibers that extend to the nucleus of Edinger-Westphal via the MLF. From this nucleus originates the motor impulse that follows cranial nerve III and the short ciliary nerve. These are responsible for the *contraction* of the *pupillary sphincter*. The fibers from the anterior quadrigeminal tubercle are ipsilateral and also contralateral to the opposite third nerve nucleus. This is how both eyes are joined by this pathway for *consensual light response*.

The pupil, besides the response to light, also responds to *accommodation* and *convergency*. The stimuli for accommodation determine the synergic contraction of the *medial rectus* muscles of both eyes and constriction of the *pupillary sphincter*. The pathway for this reflex extends from the retina to the occipital cortex. From there, it extends to the motor cortex that controls the medial rectus, the pupillary sphincter, and the ciliary branch of cranial nerve III.

Illumination of one pupil (if the patient has normal vision) produces *constriction* of the *opposite* pupil and this is known as a *consensual reflex*.

Many alterations can occur in the pupil. If there is a disorder in the *shape* of the pupil, this spincter can be circular or irregular and is known as *dyschoria*. This can occur in congenital anomalies like the *colobomas* or after surgery for cataracts. Syphilis can also produce dyschoria. The pupil can also be unequal (*anisochoria*); one of the pupils may be midriatic and/or myotic. Although the anisochoria can be congenital, the enormous value of the pupillary exam should always be kept in mind. In cases of increased intracranial pressure, cerebral edema, tumor, subdural hematomas, and other such conditions, these may induce *uncal herniation*. The herniation can *compress* the *cranial nerve III* and produce paralysis of this nerve. Paralysis of this third nerve is associated with *dilated* pupils. It should also be remembered that an aneurysm of the *internal carotid artery* (at the beginning of the posterior communicating artery) can also produce *dilated pupils*. Diabetes and ophthalmoplegic migraines are notorious for producing unilateral midriasis.

Myosis is a *decrease* in the diameter of the pupil. It can be unilateral or bilateral. There are two types of myosis, *spasmodic* or *paralytic*.

The *spasmodic myosis* is caused by any lesion that produces stimulation of the *third* nerve (from its origin, to the entire extracranial course). The *paralytic myosis* is produced by a paralysis of the *sympathetic* nerve.

By and large, the examiner tends to focus his attention to the *dilated* pupil only. Usually everyone speaks, assuming that the right pupil is the dilated one, of *anisochoria with right midriasis*, but one also should raise the question "why not anisochoria with *left* myosis?"

In cases of cervical cord trauma, if the intermedial lateral column of the cervical cord is affected, the pupil in the *affected* side is *myotic*. If the patient presents with myosis, anhidrosis, hypotension, and bradycardia (and with priapism if a male) in a case of trauma, one should and *must* think of *trauma* in the *cervical cord*. The sympathetic paralysis produces a *splanchnic (visceral) vasodilatation*. Hence, *hypotension* and command of the vagus nerve is due to *sympathetic paralysis* and thus, the *bradycardia*.

If there is any doubt whether the myosis is spasmodic or paralytic (meaning, a lesion of cranial nerve III, parasympathetic, or paralysis of the sympathetic nerve), three drugs can be used to establish the diagnosis. In the case of a cranial nerve *III* lesion, or that the myosis is *spasmodic*, *atropine* produces *dilatation* of the pupils. But, if the myosis is due to a *sympathetic* paralysis, dilatation does NOT occur. *Adrenaline* and *cocaine* exaggerate the myosis. While, if the myosis is due to sympathetic paralysis, the pupil does NOT dilate. It should be remembered that *adrenaline* is a *stimulant* of the *sympathetic* system.

Spasmodic myosis can be due to *syphilis*, and also due to toxicity with *morphine*.

Paralytic myosis (due to sympathetic damage) is, in general, *unilateral*. It can be observed with a tumor at the apex of the lung, cervical traumas, or in a disorder that affects the sympathetic nerve in its course through the brain stem. In many cases, the photomotor reflex may be *absent*. This means that the pupils do NOT contract under light stimuli. It can be observed in unilateral blindness, or in paralysis of cranial nerve III. On certain occasions, the pupils do not respond to light, but *do* respond to *accommodation*. This is known as *Argylle-Robertson syndrome*. The origin of this sign is not very clear. Many postulate that it is due to a lesion of the ophthalmic ganglion or a lesion of the pupillary fibers at the anterior quadrigeminal tubercles. Although one should always think of *syphilis* when the Argylle-Robertson pupils are present, one should remember that *chronic alcoholism* and *multiple sclerosis* may be responsible for this phenomenon. In the case of trauma to the *optic* nerve or to cranial nerve *III*, there can be a *unilateral* Argylle-Robertson pupil. In some cases, the pupils can be myotic and accompanied by an absence of the ankle and patellar reflex. This is known as *Adie pupils*.

Cranial nerve III can be affected by lesions, either partially or totally. When the lesion is *complete*, the *levator palpebral* will be involved and the patient will have *ptosis*; there will be paralysis of the superior, inferior, and medial rectus. The ocular globe will be laterally deviated because cranial nerve VI (which innervates the lateral rectus) is intact. In paralysis of cranial nerve III, there will be as paralysis of the superior, inferior, and medial rectus muscle—as well as the inferior oblique. Due to the fact that the superior oblique and the lateral rectus are intact, the ocular globe can be moved slightly lateral and downward. The pupil becomes dilated and does not respond to light or to accommodation. Total paralysis of cranial nerve III is less frequent than partial paralysis. This partial paralysis can be due to one muscle alone. Increased intracranial pressure with uncal herniation can be produced by a *subdural* and *intracerebral hematoma*, a very common cause of *third nerve palsy*. The aneurysm of the *posterior communicating artery* is notorious for producing paralysis of cranial nerve *III*. Among other causes, we should mention *opthalmoplegic migraine* and *diabetes mellitus*.

At times, when there is congenital ptosis of cranial nerve III, if the examiner asks the patient to move the jaw while opening the mouth toward the opposite side of the ptosis, he or she may *elevate* the *eyelid* that is *affected with* the *ptosis*. This is known as the *Marcus Gunn phenomenon*.

Paralysis of the *orbital muscle* is called *ophthalmoplegia*. Ophthalmoplegia can be divided into internal and external. Internal ophthalmoplegia occurs when the pupillary sphincter is paralyzed. The external, on the contrary, involves all the extrinsic muscles of the eyes. Total ophthalmoplegia means that besides paralysis of the pupillary sphincter, there is also involvement of all the muscles of the orbit.

Nuclear ophthalmoplegias are caused by lesions occurring in the nucleus of origin of cranial nerve *III,* around the aqueduct of Sylvius. Usually in this type of ophthalmoplegia, the pupillary sphincter is not involved. *Infranuclear ophthalmoplegias* are produced by lesions that affect the *oculomotor* nerve separately. Among the etiologic factors we find a tumor, abscess, trauma, aneurysm, meningitis, and syphilis.

CRANIAL NERVE IV (TROCHLEAR NERVE)

Cranial nerve IV is also known as the *trochlear* nerve, and it originates in a nucleus just below the nucleus of cranial nerve III (laterally to the aqueduct of Sylvius). From here, it runs dorsal to the aqueduct and decussates in the midline. The one on the opposite side emerges in the *posterior* portion (at each side of the midline) between the superior and inferior quadrigeminal tubercles. Then, it turns around and runs into the *lateral* aspect of the cerebral peduncle (piercing the tentorial dura mater to enter in the external wall of the cavernous sinus) to enter the orbit through the superior orbital fissure. In the orbit, it supplies the *superior oblique*.

CRANIAL NERVE V (TRIGEMINAL NERVE)

The apparent origin of cranial nerve V is located at the anterior aspect of the pons, at the junction of the brachium pontis with the pons. Two roots are identified at that level, the *motor* and the *sensory*.

The *sensory* root of the trigeminal nerve has three nuclei of origin.

1. The nucleus of the *substantia gelatinosa* that extends from the medulla oblongata to the pons.
2. The nucleus of the *locus ceruleus*, located at the lateral portion of the upper triangle of the fourth ventricle.
3. The *medial* nucleus, which is located above and somewhat posterior in the protuberal portion of the nucleus gelatinosus.

This is a rather small group of cells. From there, a very small group of fibers that follow a horizontal course in the pons integrate the medial root of the sensory portion of cranial nerve V. From the gelatinous nucleus of Rolando, there is a conglomeration of fibers that follows an ascending course in the medulla oblongata before leaving the brain stem, forming what is known as the inferior root of the sensory portion of cranial nerve V. From the locus ceruleus, there is a group of fibers that have a descending course, known as the roots of the nucleus ceruleus. All these roots converge at the superior and anterior portion of the pons to form the sensory part of the trigeminal nerve.

The *motor* root has two subnuclei, a *principal* one and an *accessory* one, located in the pons. From this area, they follow a forward course to emerge in the pons below the sensory roots. From that point on, they run toward the inner part of the petrous bone toward the cavum Meckel's. The sensitive root enters the gasserian ganglion, or ganglion of Gasser. From the anterior part of the gasserian ganglion emerge the *three* branches of the nerves that gives the names of the nerves. Since there is another fifth nerve on the opposite side, it represents the twin of the opposite side and hence, the word *gemini* and three branches hence, the name *t r i g e m i n a l* (see Figure 19-6).

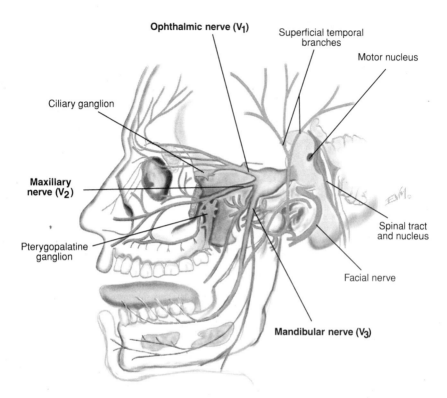

Figure 19-6 The divisions of the trigeminal nerve.

From the anterior border of the gasserian ganglion emerge these three branches:

1. The *ophthalmic division*, which enters the orbit through the superior orbital fissure. Once this nerve leaves the ganglion of Gasser, it enters in the lateral dura of the cavernous sinus, entering the orbit, dividing into three terminal branches, a lateral one—the lacrimal nerve, a medial one—the frontal nerve, and an internal one—the nasal nerve. From the nasal branch, emerges a small branch that constitutes the sensitive root of the ciliary ganglion (that becomes adnexed to the ophthalmic nerve). As stated before, the ophthalmic ganglion has a motor root that comes from cranial nerve III and a sympathetic root that originates in the sympathetic plexus—that surrounds the internal carotid at the cavernous sinus. The efferent branches of the ophthalmic ganglion constitute the *ciliary nerve* (see Figure 19-7).

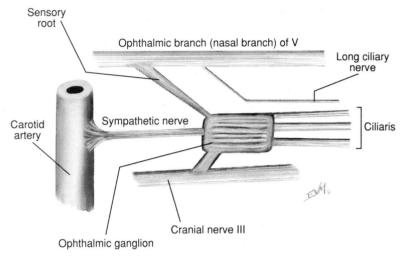

Figure 19-7 A schematic of the trigeminal nerve.

2. The *maxillary* branch, or *superior maxillary nerve*, which leaves the cranial cavity through the foramen rotundum, subsequently crosses the pterygomaxillary fossa to enter the infraorbitary canal. At the pterygomaxillary fossa, the superior maxillary nerve adnex to the sphinopalatine ganglion (sympathetic ganglion) to form the vidian nerve that subsequently runs backward to join the facial nerve. The infraorbital nerve has multiple branches that innervate the inferior eyelid, a portion of the cheek, the ala of the nose, and the upper lip. The mucosa of the superior eyelid, the hard palate, the tonsil, the uvula and nasopharynx, and the inferior portion of the nasal cavities are also supplied by this branch.
3. The *inferior maxillary nerve* is a mixed nerve that contains a *sensory* root that originates at the ganglion of Gasser, and a *motor* root that is the *masticatory nerve* itself (which represents the motor branches of the trigeminal nerve). Both nerves leave the cranial cavity through the *foramen ovale* to divide immediately into several branches. As soon as the nerve leaves the foramen ovale, the inferior maxillary nerve receives as adnex, a sympathetic ganglion of Arnold, or simply the otic ganglion. The motor branch innervates the masseteric muscle, the temporal muscle, the pterygoid muscle, the external tensor palatini, the mylohyoid, the anterior ventor of the digastric muscles, and the tensor of the tympanic membrane and malleus ossiculus. The upper part of the skin of the temporal region and the adjacent portion of the ear, anterior and superior wall of the external auditory canal, a portion of the cheek, the inferior lips, menton, the inferior teeth as well as the inferior gum, also give the sensory supply to the anterior

two-thirds of the tongue, the inner portion of the cheek, the floor of the mouth, and the inferior salivary glands. The portion that supplies the anterior two-thirds of the tongue, is a group of fibers that abandons the trunk of the inferior maxillary nerve and is incorporated to the chorda tympani, after it reaches the facial nerve. The inferior maxillary nerve has two terminal branches, the *lingual* and the *inferior dentary* nerve or *mentonian* nerve.

Two important pathologic processes can affect the trigeminal nerve:

1. Paralysis of cranial nerve V, accompanied by anesthesia of half of the face, nasal and oral mucosa, and the tongue.
2. Paralysis of the masticatory muscles.

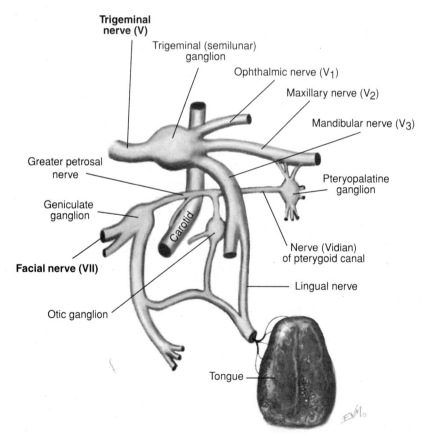

Figure 19-8 The taste pathways of cranial nerves V and VII.

This paralysis can occur in the peripheral course of the nerve, from its apparent origin in the pons. When the paralysis is central in origin, it usually affects other centers in the brain stem. A very common cause is multiple sclerosis and brain stem hemorrhage.

Trigeminal neuralgia is characterized by intense violent *pain*, of sudden onset, that occurs in a repetitive form and can affect all or, independently, each of the branches of this nerve. The most commonly affected is the *opththalmic* division, then the second division, and finally the third division. Sometimes, *trigeminal neuralgia* can produce a disorder in the trophism of the skin. A tumor of the gasserian ganglion, petrositis, diabetes, or syphilis can also be an etiologic factor in trigeminal neuralgia.

If the gasserian ganglion is affected by *herpes zoster virus*, then multiple vesicles can appear in the distribution of the affected nerve, and the patient may develop a post-herpetic *neuralgia*.

Due to the fact that the cranial nerves III, IV, VI, and the ophthalmic branch of cranial nerve V are located at the lateral wall of the cavernous sinus and traverse the superior orbital fissure (and subsequently are located in the orbital cavity), certain syndromes may affect all of these nerves *at the same time*. Among them, should be mentioned, *thrombosis* of the cavernous sinus, Rochon-Duvigneaud syndrome, and syndrome of the superior orbital fissure. The last one is usually produced by an *inflammatory* process affecting the superior orbital fissure. In the orbital cavity, the pseudogranuloma of the orbit, known as *Tolosa Hunt*, can also produce a *complete ophthalmoplegia* in the affected side, with loss of sensation in the distribution of the ophthalmic branch of cranial nerve V. Very often, it is accompanied by *exophthalmia*.

CRANIAL NERVE VI (ABDUCENT)

Cranial nerve VI (*abducent nerve*) originates in a nucleus just below the ependyma (in the floor of the fourth ventricle) that corresponds to the pons. From there, it runs a forward course traversing the pons to exit on each side of the *foramen cecum* of the pontomedullary sulcus. Cranial nerve VI is one of the largest intracranial nerves. It extends forward to the dura mater at the level of the sella turcica and enters inside the cavernous sinus. Then, it enters the orbit through the superior orbital fissure innervating the *lateral rectus muscle*.

Paralysis of cranial nerve VI produces deviation of the eyeball *toward* the *midline*. Cranial nerve VI can be affected at the tip of the petrous bone and it may be accompanied by involvement of the gasserian ganglion, and this is a syndrome known as the Gradenigo's syndrome, commonly due to ostiitis at the tip of the petrous bone.

CRANIAL NERVE VII (FACIAL NERVE)

This nerve has a *motor* portion that is the *facial nerve* itself, and a *sensory* root known as the *nervus intermedius*.

The *motor* root, as mentioned in the description of the pons, has a nucleus of origin in the pons, and a small group of cells located in the upper part of the medulla oblongata. The motor root has a serpentine course that runs posteriorly and wraps around the nucleus of cranial nerve VI, forming the *colliculus fascialis*. Subsequently, it courses laterally to exit at the supraolivary fossa.

The *sensory* nerve of the facial nerve, or *nervus intermedius*, has a nucleus of origin in the most upper part of the fasciculus solitarius, in the most superior portion of the gray ala (see Figure 19-9). From there, it runs an oblique course forward and laterally to exit also at the supraolivary fossa, between the motor portion of the facial nerve proper and the auditory nerve. From that area, it enters the subarachnoid space (in the cerebellopontine angle) to enter the internal auditory canal.

The facial nerve and the nervus intermedius soon penetrate into the fallopian aqueduct. The nevus intermedius ends in the *geniculate* ganglion. This ganglion gives a small branch, which mingles in an intimate fashion with the facial

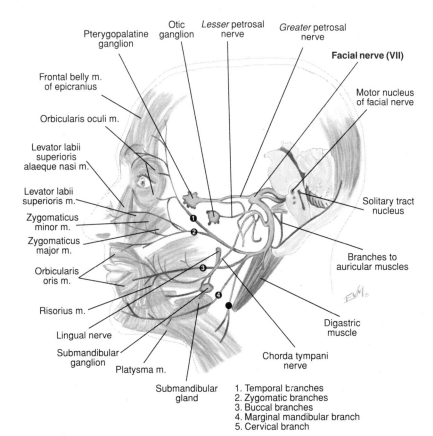

Figure 19-9 Facial nerve innervation.

nerve itself. From the geniculate ganglion on, the facial nerve is a mixed nerve. After the facial nerve traverses the fallopian aqueduct, it leaves the petrous bone through the stylomastoid foramen, and after about a distance of 1.5 cm, it enters the parotid gland. Here, the nerve is divided into two terminal branches, the *temporofacial branch*—the largest of the two branches, which is divided into three sub-branches (the temporal, malar, and infraorbital).

The *temporal* branch supplies the frontal portion of the occipitofrontalis muscle, the orbicularis palpebral, and the supraciliary muscle, and joins the lacrimal branch of the ophthalmic nerve.

The *malar* branch travels across the malar bone to supply the orbicularis palpebral.

The *infraorbital* branch gives the nerve supply to the pyramidalis nasi, the levator auris anguli, and the levator labii superioris.

The *cervicofacial division* follows a forward and downward course, crossing the external carotid artery. Exactly at this level, it receives the great auricular nerve. At the level of the inferior angle of the jaw, it gives several branches that are divided in three sets: the *inframaxillary*, the *supramaxillary*, and the *buccal* branch.

The *buccal* branch runs laterally to the masseteric muscles, and the buccal branch supplies the orbicularis auris and the buccinator muscle.

The *supramaxillary* branch runs toward the platysma and the depressor anguli auris and joins the mental branch of the inferior dentary nerve.

The *inframaxillary* cervical branch supplies the platysma muscle.

Besides these terminal branches, at the level of the fallopian aqueduct, the facial nerve gives origin to ten branches (five branches at the level of the aqueduct and five branches outside the petrous bone). The first group of branches, known as *intrapetrosal branches*, are:

1. The greater or large superficial petrosal nerve, which ends at the sphenopalatine ganglion.
2. The lesser superficial petrosal nerve, which ends in the otic ganglion. These last two nerves join the petrosal branch of the glossopharyngeal nerve.
3. The tympanic branch, which innervates the stapedius muscle.
4. The chorda tympani, which joins the lingual branch of the trigeminal nerve, a branch that anastomoses with the vagus nerve.

As the nerve exits from the stylomastoid foramen (extrapetrosal branch), the facial nerve gives branch to:

1. The glossopharyngeal nerve.
2. The posterior auricular nerve.
3. The posterior venter of the digastric muscle.
4. The stylohyoid muscle.

At the internal auditory meatus, the facial nerve gives a branch that joins the auditory nerve.

The nucleus of the facial nerve receives cortical innervation from the most inferior part of the motor cortex. From there, the axons reach the internal capsule at the level of the genu. In the exploration of the facial nerve, the examiner first has to notice the *symmetry* of the face. Look to see whether or not an eye appears more open, as compared to the other eye, and if the patient has a lot of tears in the eye that appears more open. The examiner should ask the patient to frown, to open and close the eyes, to move the commissure of the lips sideways, and to project the lower lip forward. The *facial* nerve is more frequently paralyzed than any other cranial nerve. The paralysis can be due to central causes, like tumors, cerebral hemorrhage, or there can be peripheral facial paralysis. Peripheral facial paralysis occurs when the lesion of the nerve occurs in the nucleus of origin, or when it happens along its peripheral course. In the peripheral course, paralysis can occur at the intrapetrosal portion, in the area known as *aqueduct of Fallopius*. This can occur below the geniculate ganglion, where it involves the chorda tympani. In which case (besides the paralysis of the facial muscles), the gustatory sensation in the *anterior* two-thirds of the tongue is lost. The patient will also present with painful hearing for the lower tones, known as *hyperacousia*.

If paralysis of the facial nerve occurs *between* the *supraolivary fossa* and the *geniculate ganglion*, we will find involvement of all facial muscles, but the gustatory disorder in the anterior two-thirds of the tongue will not be present (see Figure 19-10). If the lesion occurs *after* the facial nerve exits through the *stylomastoid foramen*, there will be a complete paralysis of the facial muscles known as *Bell's palsy*. The eye on the *affected* side will be more open than the unaffected side. In the same affected eye, there will be abundant *lacrimation*. If the patient attempts to close the eyes, the eyeball in the affected side tends to deviate upward and outward; this is known as *Bell's phenomenon*. If the paralysis occurs in the motor nucleus, the Bell's palsy will occur without the gustatory disorder but it will involve, concomitantly, cranial nerve VI. The so-called superior paralysis of the facial nerve is a paralysis in which only the inferior portion of the facial nerve will be affected, while the upper part of the facial nerve will suffer no alteration.

The etiology of facial nerve paralysis depends on whether the paralysis is *peripheral* or *central*. The most common causes of peripheral nerve palsy include the following.

1. *Infections* (like syphilis, otitis media, leprosy, and herpes zoster). If the herpes zoster affects the external auditory canal, the whole ear, and the facial nerve, this disorder is known as *Ramsey-Hunt syndrome*.
2. *Metabolic disorders* (like gout, diabetes, and beri-beri, etc.) In children, one of the most common causes of bilateral peripheral seventh nerve palsy is the *Guillian-Barré syndrome*. The facial nerve can also be affected in the peripheral course by lymphomas, leukemia, and/or sarcoid.
3. Trauma to the skull base affecting the petrous bone.
4. Pontine hemorrhage or tumor.

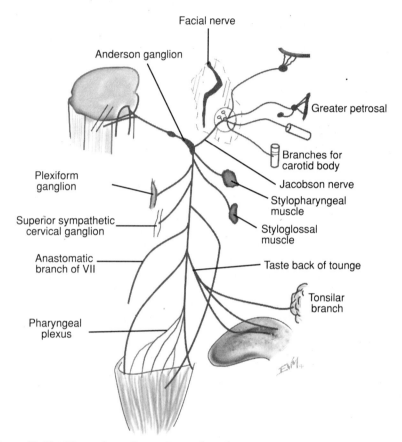

Figure 19-10 The oral ganglion and nerve branches.

In supranuclear paralysis of the facial nerve, the paralysis is always *contralateral* to the lesion. Among the most common causes are occlusion, tumor, hemorrhage, or abscesses. *Hemiparesis* or *hemiplegia* also accompanies this supranuclear paralysis.

The facial nerve can be affected by a vascular disorder, *hemifacial spasm* —where blocked contraction of the muscles of the face occurs on the *affected* side, which becomes more pronounced when the patient is tense or excited.

CRANIAL NERVE VIII (AUDITORY NERVE OR VESTIBULAR-COCHLEAR NERVE)

The auditory nerve has the exclusive function of *hearing*. It is distributed in the internal ear, and also intervenes *equilibrium*. Cranial nerve VIII has two sets of nerve fibers and is concerned with the transmission of hearing from the inner ear to the pons and medulla oblongata, and from there to the brain and cerebellum. The *acoustic* nerve runs from the supraolivary fossa to the internal auditory canal. Once it reaches the internal auditory canal, the acoustic nerve is divided in two terminal branches. The *anterior* branch, which is the *cochlear nerve*, and a *posterior* branch, which is the *vestibular nerve*. The vestibular nerve joins the cells of the ganglion of Scarpa.

The vestibular nerve is subdivided into three branches: a *superior* branch (that goes to the utriculus and to the external semicircular canal), an *inferior* branch (that goes to the sacculus), and a *posterior* branch (that extends to the posterior semicircular canal).

The fibers that integrate the vestibular nerve, once they reach the fourth ventricle, are divided into two branches (an *ascending* and a *descending branch*).

The *ascending* branch ends in three nuclei: the *vestibular* nucleus (also known as the nucleus of Deiters), some fibers that run toward the flocculonodular system following the inferior cerebellar peduncle (which originate from the internal dorsal nucleus), and other fibers, which end at the nucleus of Bechterew (in the most dorsal portion and lateral part of the medulla oblongata).

The *descending* fibers (also known as the *inferior root*) can be followed as far down as the cervical cord, and are joined together to a group of cells known as the *descending vestibular fibers* and *nucleus*. The most important pathway of the vestibular system, besides the one that runs toward the cerebellum, is the fibers of the *medial longitudinal fasciculus*. In this fasciculus, there are ascending and descending vestibular fibers establishing connections with the nuclei of cranial nerves III, IV, V, VI, IX, and X on both sides. From the terminal nucleus of the vestibular nerve, there are fibers that run transversely and toward the midline. They cross with the one on the opposite side, reaching the reticular formation of the pons and medulla oblongata, and become "verticalized." Then, they join the *medial lemniscus* to establish connection with the cerebral cortex at the temporal lobe.

The *cochlear nerve*, or *dorsal root* of the *acoustic nerve*, is located external to the vestibular root. It ends in two nuclei. The *accessory* nucleus is located in the floor of the fourth ventricle, immediately medial and anterior to the restiform body. The other branch is located laterally. From the bipolar ganglion of Corti, the axons join together to form the auditory fibers that enter the brain stem at the lowest portion of the fourth ventricle (at the *pontine* portion of the fourth ventricle). A few of these auditory fibers terminate in the superior olivary nucleus and in the nucleus of the corpus trapezoid. From this nucleus, certain fibers decussate at the trapezoid body to *ascend* in the *lateral lemniscus*, to the inferior colliculi and medial geniculate body. Along the retrolenticular portion of the internal capsule, the auditory fibers reach the superior temporal gyrus. The fibers that depart from the lateral acoustic nuclei run in the floor of the fourth ventricle and they are seen with the naked eye, integrating with what is known as the *stria acousticus* (which end in the *superior* olive of the pons and some decussate to the opposite side).

The cochlear nerve conducts all the *auditory* sensations that are received by the organ of Corti. The *vestibular* nerves carry with them the specific fibers that deal with *equilibrium* and *orientation* (revealing the different positions of the body in relation with the space). The *semicicular canal* is stimulated by active movements of the head. To explore this nerve, it is necessary to address the examination independently to each one of the two nerves that integrate the auditory nerve, the cochlea and vestibular nerve. Functional hearing tests are normally performed by an audiologist. However, the clinician can perform certain exams during a regular examination.

For example, a *Weber test* is a simple test and can be done with a tuning fork. The tuning fork is placed in the *middle* of the cranial convexity and the patient is asked to tell us how he or she perceives the vibration—equal in both ears, or better on one side as compared to the other side. For example, if the patient does hear better in the right ear, we say that the Weber is *lateralized toward that side*.

In the *Rinne test*, the vibrating tuning fork is placed at the *apex* of the *mastoid* and the patient is asked to inform the examiner when he or she *ceases* to perceive the sound. Then, quickly with the tuning fork that is *still* vibrating, the fork is placed in front and lateral to the ear. Normally, the patient should *hear* the vibrating tuning fork, and one then speaks about a *positive* Rinne test. If, on the other hand, the patient does NOT hear the sound, we say that the Rinne test is *negative*. This indicates that there is better *bone* transmission than air transmission. The cochlear nerve may sustain a lesion in the inner ear, in the *intrapetrosal* portion as well as in the cortical center. There are several types of *deafness*: *nerve deafness* that is due to a lesion either in the acoustic nerve or in the cochlear system; deafness due to sclerosis of the middle ear bone; senile deafness; and finally, deafness produced by a lesion of the central nervous system.

The examination of the *vestibular system* is normally carried out by an ear, nose, and throat specialist. The vestibular semicircular canal, which is concerned with the posture and equilibrium, consists of three semicircular canals, the *sacculus*, and the *utricle*. There are three semicircular canals for each side, the external canal, the anterior or horizontal canal, vertical canal and the posterior vertical canal. Within this system of the semicircular canal, circulates the *endolymphatic fluid*. Movement of the endolymphatic fluid is the result of movement of the head. Irritation of the *vestibular* branch of the acoustic nerve can either produce *vertigo* or *buzzing noises*. Many times, vertigo is associated with a sensation of nausea, vomiting, buzzing of the ear, diaphoresis, and skin pallor. Alteration of this vestibular system can produce *nystagmus*. The nystagmus can be classified as central or peripheral. The are certain general characteristics that help the examiner to determine if the nystagmus is central or peripheral.

Peripheral nystagmus should be understood as the one that occurs as a consequence of a disruption of the *vestibular* system and apparatus from the *semicircular canal* to the point of entrance of the nerve into the brain stem.

Central nystagmus occurs when the lesion involves the *nuclei* of the *vestibular nerve* at: the brain stem, a lesion of the brain stem, the flocculonodular system and cerebellum, thalamus, frontal lobe, and frontoparietal junction.

Peripheral nystagmus is transient, and central nystagmus tends to be permanent. Peripheral nystagmus has to be elicited, and central is spontaneous. Peripheral nystagmus is normally positional, and central nystagmus *can* sometimes be positional but *paroxysmal*. This means that if it occurs in only one position, in *paroxysmal*, there is a possible lesion in the vestibular nucleus of the fourth ventricle. If the patient is from an underdeveloped country, one should think of the presence of a cysticercus irritating the vestibular nucleus of the fourth ventricle. This is called *Nylen nystagmus*. If the nystagmus occurs in vertical, horizontal, and/or rotatory fashion, it is called *pervertic nystagmus*, or *Nylen two*. In peripheral nystagmus, there can be vestibulocochlear dissociation; the cochlear branch can be spared. In central nystagmus, vestibulocochlear dissociation is rare. In nystagmus of cortical origin, the abnormal eye movements are of *equal* excortion; they have the appearance of pendular movement. Notice that nystagmus of *vestibular* origin has two associated ocular movements; the first is *slow*, and the second is *rapid*. The eyes return abruptly to the neutral position of the initial movement. The *slow* component is the most fundamental element of vestibular nystagmus; whereas, the *rapid* or accessory component is purely reactional. A lesion, for example, that affects the labyrinth system on the *left* will produce nystagmus with a *slow* component toward the *left*.

Peripheral Nystagmus	Central Nystagmus
Disruption of the vestibular system.	Disruption of the nuclei of the vestibular nerve
Transient nystagmus	Permanent nystagmus
Has to be elicited	Spontaneous
Positional	Sometimes positional, but *paroxysmal*
Vestibulocochlear dissociation	Vestibulocochlear dissociation is rare

The neuro-otologist, by and large, is the one that performs the *caloric test*. The reader is directed to study and investigate in a neuro-otology book, the technique and results of the caloric test. Among the etiologic factors that affect the labyrinth, is *labyrinthitis* in *Meniere's syndrome*. In the central nervous system, possible etiologies of *vertigo* include: a tumor that affects the nucleus of the fourth ventricle, cerebellopontine angle tumor, multiple sclerosis, and lesions in the cerebellum and cerebral peduncle.

Nystagmus is very common in *cerebellar* lesions and usually occurs *toward* the side of the lesion. In addition, certain toxic elements like alcohol, barbiturates, and Dilantin (in toxic levels) can create nystagmus. A tumor can affect the *acoustic* nerve. This may occur in young patients and is usually associated with *neurofibromatosis multiplex*. This tumor originates primarily from the *vestibular nerve*, beginning in the *internal auditory meatus*, and it may grow to invade the entire cerebellopontine angle.

The symptoms of an *acoustic neuroma*, in general, occur over many years. This tumor tends to affect or paralyze the *facial* nerve, and according to the size may produce dysarthria, dysphagia, and hypalgesia in part of the face and in the cornea.

CRANIAL NERVE IX (GLOSSOPHARYNGEAL NERVE)

The glossopharyngeal nerve is a *mixed* nerve that contains sensory, secretory, vasomotor, as well as motor fibers.

The *motor* fibers innervate the stylopharyngeal muscle and have a motor component that joins the vagal nerve as well as the sympathetic nerve. These motor fibers give a nerve supply to the muscles of the pharynx, tonsil, and soft palate.

The *sensory* fibers perceive sensory impulses from the mucosa of the pharynx and middle ear, as well as gustatory impulses from the back of the tongue. As in all the cranial nerves, the glossopharyngeal nerve has a nucleus of *origin* and *apparent origin*—which is the point of exit from the central nervous system. The *apparent* origin consists of several filaments that exit in the upper part of the medulla oblongata (between the olive of the medulla oblongata and the restiform body). From this area, the nerve (which is located above the vagus nerve) runs upward and forward and exits the skull through the uvular foramen. This is between the jugular vein and internal carotid artery. Then, it descends in front of the carotid to curve inward, lying between the stylopharyngeal muscle and the middle constrictor of the pharynx. It is located below the hypoglossal nerve to distribute in the mucosa of the tongue, tonsil, and mucous glands of the mouth. Once it passes through the uvular foramen, the nerve has two enlargements. One is *superior*, and known as the *jugular ganglion*, which is small and is situated in the most adjacent portion of the nerve to the jugular foramen. There is an *inferior ganglion*, which is *larger* and is known as the *petrous ganglion of Andersch*. From this ganglion, arise several fibers that connect cranial nerve IX with the vagus and sympathetic nerve. The branch to the vagus nerve also arises from the inferior ganglion. The sympathetic branch connects the superior cervical ganglion to the inferior petrosal ganglion. There is a branch that pierces the posterior center of the *digastric muscle* that also originates at the petrous ganglion and joins the facial nerve immediately after the facial nerve exits through the stylomastoid foramen.

The glossopharyngeal nerve has a branch from the inferior petrous ganglion that follows an upward and backward course. This is known as the *tympanic branch*, or the *nerve of Jacobson*, which enters a rather small canal in the lowest surface of the petrous bone (located between the jugular fossa and the carotid canal). It ascends into the middle ear, in the floor, and divides into branches that run in the outer surface of the promontary integrating the plexus tympanicus. From this plexus originates the *lesser superficial petrosal nerve*, and another branch that joins the greater superficial petrosal nerve. And then, a third branch that distributes in the mucosa of the middle ear.

The *carotid branches* of cranial nerve IX descend along the *carotid artery* to join branches of cranial nerve IX and the sympathetic nerve. The *lingual* branches, in sets of two, distribute over the papilla of the mucosa of the tongue and other small branches that join the lingual nerve. The nerve of Jacobson carries the *secretory* as well as the vasomotor fibers, and the secretory fibers are distributed in the parotid gland. Remember, the *gustatory* fibers at the base of the tongue originate at the glossopharyngeal nerve, the lingual nerve, and the chorda tympani branch of the facial nerve—which originates at the geniculate ganglion and provides the gustatory perceptors for the anterior two-thirds of the tongue. All the gustatory impulses for the lingual nerve follow the *chorda tympani* to the geniculate ganglion and the nervus intermedius, to end at the fasciculus tractus solitarius. This is where fibers follow an ascending course and join the medial lemniscus to reach the hippocampus.

To explore the gustatory function of cranial nerve IX, we should remember that the *back* of the tongue is the gustatory territory of cranial nerve IX. Remember, the **b**ack of the tongue perceives **b**itter and the *anterior* two-thirds perceives *sweet and sour*. The lack of taste is called *agustia*. The motor exploration of the glossopharyngeal is in reality a compound exploration of IX and X, since these two nerves innervate the *constrictor* and *levator* of the pharynx; therefore, when there is difficulty in *swallowing* we are really saying that these two nerves are affected. There is also lack of a *gag reflex*. The glossopharyngeal nerve can present a special type of *neuralgia* known as the *morbus of Harris*—which is characterized by episodes of sharp pain in the affected side of the pharynx. Initially, the pain is very intense, can last seconds, seldom minutes, and it comes and goes. It is always *unilateral*. Sometimes this can be due to carcinoma of the pharynx, and at other times, the etiology is unknown.

The real origin (or nucleus of origin) of cranial nerve IX has two nuclei. The origin is located at the level of the *gray ala*, and is known as *nucleus of the gray ala* and the *fasciculus tractus solitarius*. Both are located in the floor of the fourth ventricle. These two represent the *sensory* nucleus of cranial nerve IX. The *motor* nuclei originate at the *nucleus ambiguous*—which is also the origin of the *motor* fibers of the *vagus* and *spinal accessory* nerves.

CRANIAL NERVE X (VAGUS OR PNEUMOGASTRIC NERVE)

The vagus nerve is a complex cranial nerve. The vagus nerve has a very extensive distribution. It traverses the neck, thorax, and the upper abdomen. It contains *motor* as well as *sensory* fibers. It supplies the vocal cords, lungs, heart, pharynx, esophagus, and stomach with motor fibers. The *apparent origin* takes place in a groove located between the olive of the medulla oblongata and the restiform bodies. It is located just below the glossopharyngeal. Along its course, it has two superimposed ganglions, one *superior* also known as *jugular ganglion* (at the jugular foramen), and another *inferior* also known as *flexiform ganglion* (which is located 1 cm below the previous one).

During its course at the neck, the vagus nerve anastomoses with the facial nerve, the glossopharyngeal, the hypoglossal, and the sympathetic nerve. At the level of the *plexiform ganglion*, cranial nerve XI integrates with the vagus nerve, for which reason this is called the *vagospinal nerve*. In the neck, the nerve is located within the sheath of the carotid vessel between the internal carotid and uvular veins; and lower down, it is located between the same jugular vein and the common carotid artery.

On the *right* side, the nerve passes across the subclavian artery, between this artery and the innominate vein, and descends laterally to the trachea, then, to the back of the pedicle of the lungs—where it spreads and forms a network known as *posterior pulmonary plexus*. From this area, it descends to the esophagus, where it divides into branches that join the vagus from the opposite side, to form a plexus-like structure known as the *esophageal plexus*. Below these branches, it forms a single cord that is distributed in the posterior surface of the stomach and joins the left side of the plexus. It sends branches to the splenic plexus and to the celiac.

On the *left* side, the vagus nerve enters just below the left carotid and subclavian arteries (behind the innominate vein). It then crosses the arch of the aorta and descends behind the pedicles of the lungs forming the *posterior pulmonary plexus*. It joins through the branches to the nerve of the *right* side. The vagus then distributes in the anterior surface of the stomach and lesser curvature, from which there are a number of nerve fibers that enter the gastrohepatic omentum. At the level of the jugular fossa, the vagus nerve gives origin to the meningeal and auricular branches. In the neck originate: branches to the pharynx, the superior laryngeal nerve, and the recurrent laryngeal and the cervicocardiac branches. In the thorax, we have: the anterior and posterior pulmonary branches, cardiac branches, and the esophageal branches. And in the abdomen, there are fibers to the stomach.

As the nerve emerges from the jugular foramen, it gives origin to *recurrent branches* that are distributed in the dura mater of the posterior fossa. The auricular branch, shortly after its origin, joins a branch of the glossopharyngeal nerve and runs lateral to the jugular vein. It then traverses the temporal bone, and crosses the fallopian aqueduct (close to the stylomastoid foramen). This is where an ascending branch joins the facial nerve, and subsequently joins the posterior auricular nerve.

The principal *motor* branch to the pharynx arises from the *inferior ganglion* of the vagus nerve. It consists of small filaments that are in reality a portion of the *spinal accessory*. They extend to the upper border of the medial constrictor of the pharynx. This is where the glossopharyngeal nerve, the superior pharyngeal nerve, the sympathetic, and the hypoglossal nerve join.

The *superior laryngeal nerve* provides sensation for the mucosa of the *larynx*. It originates at the *inferior* ganglion of the vagus, also is known as *ganglion plexiform*. The superior laryngeal nerve runs medially and joins with filaments of the superior sympathetic cervical ganglion. It then descends laterally and externally to the larynx and is divided in the external laryngeal branch that supplies the cricothyroid muscle. It also communicates with the superior cardiac nerve behind the carotid artery. The *internal* branch of the *superior laryngeal* nerve descends to the level of the thyrohyoid membrane, passing together with the superior laryngeal artery, and distributes in the mucosa of the larynx, specifically the epiglottis and the base of the tongue. This is where others come backward to supply the mucous membrane of the superior orifice of the larynx and vocal cords. There are small branches that join the recurrent laryngeal nerve at the inner surface of the thyroid cartilage.

The *inferior laryngeal nerve* is also known as the *recurrent laryngeal nerve* because its course turns upward. It originates at the *right* side, in front of the subclavian artery, and runs behind those vessels. It ascends medially to the side of the trachea (behind the common carotid artery) and in front of the inferior thyroid artery. In the *left* side, it arises in *front* of the *aortic arch*, runs a backward course (where the ductus arteriosus connects the pulmonary artery with the aortic arch), and ascends lateral to the trachea. Both nerves are located in between the trachea and the esophagus. At the lower border of the inferior constrictor of the pharynx, they enter the larynx and distribute to all the muscles of the larynx (except the cricothyroid). The nerves communicate and form the *superior laryngeal nerve*.

At the cervical level (upper and lower part of the neck) originates the superior branch that joins the sympathetic cardiac branch. This can be followed all the way to the deep cardiac plexus. The inferior cervical branch takes origin just above the first rib. On the *right* side, it runs in front of the innominate artery. On the *left* side, it passes in front of the aortic arch to join the superficial cardiac plexus. At the level of the *thorax*, the vagus nerve gives origin to the *cardiac* branches (anterior pulmonary branches that join the pulmonary plexus). The pulmonary branches are divided into anterior and posterior pulmonary branches. The *anterior* pulmonary branches join the filament of the sympathetic nerve to form the anterior pulmonary plexus. The *posterior* branches also join the sympathetic branches and fibers from first, second, and third thoracic sympathetic ganglion to integrate the posterior pulmonary plexus.

The branches to the esophagus originate *below* the pulmonary branches and join on the opposite side to form the *esophageal plexus*.

The *gastric* branches are the *terminal* filaments of the pneumogastric or vagus nerve. The branches from the *right* vagus distribute in the posterior surface of the stomach and join the *celiac* plexus. The branches from the *left* vagus nerve are distributed in the anterior surface of the stomach and lesser curvature. Sensory fibers from the vagus nerve transmit chemoreceptor impulses from the aortic glomus, and with cranial nerve IX, from the carotid sinus. These fibers end in the *solitary nucleus*.

The vagal nerve, from its origin, is a mixed nerve with motor, sense, and regulation of vegetative functions. The vagus nerve has motor nerve functions in charge of the striate fibers of the pharynx and larynx, the smooth fibers from the bronchi, the esophagus, stomach, and small bowel; moderating fibers to the activity of the heart muscle; and secretory functions of the stomach, pancreas, and trachea. To study cranial nerve X in a patient, one has to study the muscles of the soft palate and the larynx. One should do pharmacodynamic testing to evaluate this complex component of the parasympathetic system.

For example, if a paralysis of the soft palate exists, an examiner can observe that the half of the *soft palate* that is NOT affected moves upward and *toward* the *healthy* side. If the patient is asked to drink fluid, the fluid may come through the nose. When the patient speaks, there may be a twining and raspy voice. In *bilateral paralysis*, there is immobility of both vocal cords with cardiac and respiratory abnormalities that compromise the survival of the patient.

CRANIAL NERVE XI (SPINAL ACCESSORY NERVE)

Cranial nerve XI is also known as the *accessory nerve of the vagus nerve*. This nerve has fibers of bulbar origin, located in the most inferior portion of the nucleus ambiguous (at the medulla oblongata). It also has some medullary nuclei, which occupy the anterior and external portion of the spinal cord (just above the fibers of origin of the first cervical nerve) and extend upward to the inferior nucleus of the fifth. This group of fibers joins together and exits the brain stem. It runs toward the jugular foramen. After it passes the jugular foramen, it divides into two terminal branches. The *internal* branch quickly ends at the nucleus plexiform of the vagus nerve, where they are completely fused. The *external* branch distributes through the *trapezius* and the *sternocleidomastoid muscle* (SCM). This nerve has the main function of moving these two muscles, the superior constrictor of the pharynx and all the muscles of the larynx (except the cricothyroid). When this nerve is paralyzed, the sternocleidomastoid does not contract in the rotation of the head towards the opposite shoulder. In a long-standing paralysis, the shoulders fall forward and the scapula pulls down and outward. There are many pathologies that can affect this nerve, including: trauma, a posterior fossa tumor, and a surgical procedure. Paralysis of central origin is seen in *amyotrophic lateral sclerosis* (ALS) and in *polio*.

CRANIAL NERVE XII (HYPOGLOSSAL NERVE)

This nerve is strictly for *motor* function. The nuclei of *origin* are two in number and both are located in the medullary portion of the floor of the fourth ventricle. The principal one corresponds to the *white internal ala* of the fourth ventricle, and the *accessory* is in front and somewhat lateral to the main nucleus. From this nucleus, there is a conglomeration of fibers that runs a lateral and forward course to exit at the preolivary sulcus. They are joined together in a solid trunk that goes through the anterior condylar foramen. Then, after a long course in the lateral portion of the neck, they distribute in the musculature of the tongue innervating all the muscles of the *tongue*.

During its course, the hypoglossal nerve receives anastomoses from the sympathetic, from the vagus, and the two first cervical nerves. It also emits a very important collateral, known as the *descending branch of the hypoglossal*. The descending branch anastomoses with the deep cervical plexus forming the *ansa hypoglossi*. From this ansa, originates the innervation to the homohyoid, thyrohyoid, and genihyoid muscles.

In the paralysis of the hypoglossal nerve, one will observe *hemiatrophy of the tongue*. If the patient is asked to stick out the tongue and to move the tongue from side to side, it will be easy to detect the paralysis. When the patient sticks out the tongue, the tip of the tongue is deviated *toward* the *paralyzed* side—due to the contraction of the healthy genioglossus muscle.

The Meninges 20

The meninges are membranes that cover the brain, the cerebellum, and the spinal cord. There are three meninges: the dura mater, the arachnoid membrane, and the pia mater.

DURA MATER

The dura mater is a thick membrane, hard and nonelastic ("durable"), which lines the interior aspect of the skull. The internal portion is shiny and glistening; the outer surface is rough. As a person ages, it adheres to the inner surface of the cranial bone. In fact, the dura mater represents the *inner periosteum* of the cranial bone. From the inner surface of the dura mater, arise four septi.

1. The *falx cerebri*—which separates the two cerebral hemispheres.
2. The *tentorium* (with two leaves, *right* and *left*).
3. The *falx cerebelli*.

It also sends, at the base of the skull, a tubuler prolongation that surrounds the nerve that emerges through the various foramina of the skull. This forms a sheath or membrane through and around the nerve that emerges through the previously mentioned foramina.

At the margin of the *foramen magnum*, the dura appears to be very adherant to the bone and continues with the dura mater that surrounds the spinal cord in the spinal canal. In the outer part of the dura mater, along the longitudinal sinus, there are the *pacchionian granulations*. The dura mater at the brain has various venous channels, which are located in the separation of the two layers of the dura. These are called *sinuses* of the dura mater. There are a total of fourteen sinuses. These are:

1. The superior longitudinal sinus.
2. The inferior longitudinal sinus.
3. The straight sinus.
4. The lateral sinuses.

5. The occipital sinuses.
6. The cavernous sinuses.
7. The circular sinus, anterior and posterior.
8. Superior petrosal sinuses.
9. Inferior petrosal sinuses.
10. The transverse sinuses.

The *superior longitudinal sinuses* are located along the sagittal sutures and occupy the upper part of the falx cerebri. They extend from the foramen cecum (at the base of the skull) to finish at the *turcular herophili*. This sinus is triangular in shape with the vertex inferior and the base close to the inner table of the sagittal suture. Along its course, it has several openings that correspond to the termination of the different cortical veins that end at the superior sagittal sinus.

The *turcula* is a venous confluency of the superior and lateral sinuses. It is located in front of the *inion*. The occipital sinus also ends at the level of the torcula. The inferior longitudinal sinus is in the free margin of the falx cerebri, in the posterior two-thirds. It also receives small cortical veins. The straight sinus is located at the junction of the tentorium with the falx cerebri. Several veins of the inferior aspect of the temporal and occipital lobe end there. The inferior sagittal sinus ends in the most anterior portion of the straight sinus. The vein of Galen and the superior cerebellar vein end at the origin of the *straight* sinus (anteriorly). The lateral sinuses are located at the most posterior margin of the tentorium; they begin at the torcula herophili, and run transversely outward toward the base of the petrous bone. They curve downward to reach the jugular foramen to continue with the internal jugular vein. Several veins of the epicranium drain into the lateral sinus to a small foramen of the skull. The occipital sinus is very small and is located at the most posterior portion of the falx cerebelli as this dural membrane attaches to the inner table of the occipital bone. The circular sinuses are two dural venous channels, anterior and posterior, and they connect the two cavernous sinuses; one passes behind and other in front of the pituitary gland. The superior petrosal sinus is located at the superior border of the petrous bone and it communicates the cavernous sinuses with the transverse sinus. The cavernous sinuses, two in number, are located one on each side of the sella turcica and extend from the superior orbital fissure to the tip of the petrous bone. Anteriorly, they receive the opthalmic vein.

In the interior of each cavernous sinus is the internal carotid artery, which is surrounded by sympathetic plexuses. Lateral to the carotid artery but, in intimal relation, is cranial nerve *VI*. In the outer wall of the cavernous sinus, from superior to inferior, are cranial nerves III, IV, and the opthalmic branch of V.

The sphenoparietal sinus, which runs along the posterior border of the lesser wing of the sphenoid, ends in the cavernous sinus. The cavernous sinus communicates with the lateral sinus to the superior and inferior petrosal sinuses.

The *inferior petrosal* sinus is situated in a groove located between the posterior border of the petrous bone and with the basilar portion of the occipital bone. Veins from the *inner ear* drain into the inferior petrosal sinuses.

The *transverse* sinus consists of several interconnected veins, located between the two layers of the dura mater, over the basilar canal of the occipital bone.

The dura mater has a conglomeration of fibers incompletely separated. The *falx cerebri* has a sickle shape with the vertex anterior and the base resting over the tentorium. The falx cerebri *separates the two cerebral hemispheres*. Its open margin is convex and contains the previously described *superior sagittal* sinus. Its lower border is free and contains the *inferior longitudinal* sinus.

The *tentorium* is a lamina of the dura mater, located between the brain and the cerebellum. It covers the upper surface of the cerebellum. Posteriorly, it attaches to the bone ridge of the inner surface of the occipital bone, and at this level contains the *lateral* sinuses. In the anterior and lateral portion, it extends from the superior margin of the petrous bones (containing at this level the superior petrosal sinuses). The tentorium has a free anterior border known as the tentorial opening through which courses the upper brain stem. The tentorium is attached to the *anterior clinoid*. The *falx cerebelli* is a rather small, triangular-shaped dura mater that is located right over the vermis of the cerebellum. As it runs down toward the foramen magnum, it divides into two layers that join the dura of the base of the skull on each side of the foramen magnum. The *diaphragm sella* is another horizontal portion of the dura that has an outer circular border and is continuous with the dura of the cavernous sinuses. The inner border forms a circular opening through which passes the *pituitary stalk*.

ARACHNOID MEMBRANE

The *arachnoid membrane* is a very thin membrane that envelops the brain and is located between the *pia mater* **in**ternally and the *dura mater* **ex**ternally. In between the dura and the arachnoid membrane is the *subdural space*. In the convexity of the brain, the arachnoid is transparent and passes over the convolution without going deep into the sulci of the brain. Conversely, at the base of the brain, the arachnoid membrane is *thicker* and it covers the mesial aspect of the frontal and temporal lobes. In between the two temporal lobes and the pons there is a large interval, which is known as the *cistern of the pons*. Between the undersurface of the cerebellum and the cerebellar hemisphere, there is a large subarachnoid space that extends toward the medulla oblongata and is known as *cisterna magna*.

The *pontine cistern* has a wide communication with the cisterna magna. The arachnoid membrane surrounds all the cranial nerves that arise from the brain. It envelopes them all the way down to the exit point from the skull. The subarachnoid space, in reality, is not a "space" because it has a conglomeration of trabecula that travels from the pia to the arachnoid membrane. There is another large subarachnoid space located in the upper surface of the corpus callosum, and it contains the two anterior cerebral arteries. At the level of the sylvian fissure there is a large subarachnoid space that contains the middle cerebral vein and the middle cerebral artery. The subarachnoid space, posteriorly, communicates with the fourth ventricle through the foramen of Mangendie in the middle and the foramen of Luschka laterally. The cerebrospinal fluid fills the subarachnoid space.

PIA MATER

The *pia mater* is a vascular membrane that derives its blood from the *vertebral arteries* as well as from the *internal carotid artery*. It covers the entire surface of the brain, folding down into the different convolutions and sulci of the brain. It also prolongs into the interior ventricular system forming the *cavum velum interpositum* and the *choroid plexus* of the *fourth* and *lateral ventricles*. At the base of the brain, at the anterior and posterior perforated space, there are a number of straight blood vessels that pierce the pia to enter the brain substance. The pia mater and the visceral sheath of the arachnoid membrane, together, are known as *leptomeninges*. The inner surface of the dura mater is covered by the *parietal*, or *external*, arachnoid membrane. This is known as *paquimeninges*.

Inflammation of the *leptomeninges* is commonly known as *meningitis*. Meningitis an entity that has multiple etiologic factors. It has several cardinal signs as follows.

1. Headaches
2. Vomiting
3. Intolerance to light
4. Seizures
5. Muscular contractures, particularly the nuchal muscles

Sometimes, it is impossible to flex the head, and other times, the contraction is so severe that the head is arched backwards creating the *opistotomus*. On other occasions, there are contractures of the flexor muscles of the lower extremities. The contracture of the flexor muscles of the lower extremities can be explored in two ways.

First, the *Kernig's sign*, which is explored when the patient is flat on his or her back. If the examiner attempts to passively raise the back while holding the knees of the patient down; attempting to sit the patient up, the patient *flexes* his or her knees.

1. A second way to explore the Kernig's sign (also with the patient lying on his or her back) is for the examiner to raise one of the patient's lower extremities, holding the leg at the heel. At this point, the examiner will observe that the patient is *unable to maintain the extension* of the leg and he or she will *flex* it at the knee level.
2. Another sign is the *Brudzinski sign*. This sign is also explored when the patient is flat on his or her back. If the examiner *flexes* the patient's *head* with one hand, while the other hand is holding the chest, the patient suffering from meningitis will also *flex his or her legs* at the *knee level*.

In severe meningeal infections, it is possible to find other clinical manifestations such as hemiplegia, ophthalmoplegia, and *delirium*.

Spinal puncture establishes the *definitive diagnosis* by examining the chemical component, the cell count, and culture and sensitivity of the spinal fluid. By and large, the fluid can be opalescent or purulent, the proteins might be *elevated* and the glucose *decreased*, and the cellular count of the cerebrospinal fluid reveals an *increased* white count.

In *viral* meningitis, usually, the blood sugar is relatively *elevated* in comparison with the serum blood glucose level. The etiologies of meningitis are multiple, among them is meningococcus, pneumococcus, staphylococcus, and streptococcus.

Tuberculous meningitis is usually a slow course. The common viruses that produce meningitis or meningoencephalitis are: measles, varicella, and herpes. In underdeveloped countries, *cysticercosis* could be prevalent.

Chemical meningitis can occur with: irritation of the meninges by blood, a ruptured aneurysm, and/or arteriovenous malformation. Metastasis and carcinomatosis of the meninges could also produce meningeal irritation.

The Spinal Cord 21

The spinal cord is a cylindrical-shaped neural structure located within the spinal canal. It begins at the level of the *foramen magnum* and extends up to the level of **L1**. The spinal cord does NOT occupy the entire spinal canal. At the level of the first lumbar vertebra, it ends in a filament of gray substance known *as filum terminalis*. It is surrounded by the meninges—ligaments attach the spinal cord to this membrane. The spinal cord presents different diameters. The spinal cord has two enlargements, one at the level of the *cervical column* and the other one, known as *lumbar enlargement*.

The *cervical enlargement* extends from the level of the third *cervical* vertebra to the second *thoracic* vertebral body and corresponds to the origin of the *brachial plexus*. The *lumbar enlargement* is located at the last three dorsal vertebrae and corresponds to the origin of the sacrolumbar plexus. Just below the lumbar enlargement, the spinal cord tapers down to take the shape of a cone, known as *conus medullaris*. The inferior portion of the conus is the one that continues with the *filum terminalis*. The spinal cord is somewhat flattened anterior and posterior. This flattening is more notorious at the cervical cord and the lumbar enlargement. The average length of the spinal cord fluctuates between 42 and 45 cm, and it should be pointed out that the length does NOT directly vary with the height of the person.

The spinal cord has various fissures or sulci: the *anterior and medial fissure* and the *dorsal and medial sulcus*. These fissures or sulci penetrate through a great portion of the thickness of the cord and are divided into two symmetrical halves. The *anterior and medial fissure* is somewhat shallower than the posterior one, and penetrates the spinal cord about one-third of the thickness of the cord. The bottom of that fissure represents the most anterior portion of the *anterior white commissure*. The *posterior and medial sulcus*, also known as the *dorsomedial sulcus*, is so interrupted along its course by cross-connective tissue and multiple blood vessels that one can hardly call it a fissure. This sulcus, however, extends into the cord about one-half of the depth of the thickness of the cord. The bottom of this fissure stops at the posterior border of the gray commissure.

On each side of the dorsomedial sulcus are two fissures, one on the left and one on the right, where the posterior root of the spinal nerve makes its entrance. In between the dorsomedial fissure and the posterolateral fissure exists another trench dividing this portion of the spinal cord into two tracts, the *posteromedial* and the *posterolateral columns*. In the anterior aspect of the spinal cord are found the anterolateral fissures, one for each side. At this fissure, the anterior motor roots of the spinal cord exit.

Each half of the spinal cord is divided into *anterior columns*, *lateral columns*, *posterior columns*, and *posteromedial columns*. In a transverse section of the spinal cord, you can recognize that the cord has two distinct portions, the *gray* portion and the *white* portion. The *white matter* is situated *external* to the gray matter as it integrates the major part of the spinal cord. The *gray matter* is a conglomeration of neurons forming, anteriorly, an enlargement known as the *anterior horn*, and posteriorly as the *posterior horn*.

The *anterior horn* has three major nuclear columns.

1. The *ventrolateral nucleus*, which is present from **C4** to **C8**, and at the lumbosacral level from **L2** to **S2**.
2. The *dorsolateral nucleus* of the anterior horn, which extends from the first *thoracic* level to the fourth thoracic level (**T1** to **T4**).
3. The *retrodorsal nucleus* is a large group of *motor* cells located at the **C8** to **T1** level.

The anterior horn and posterior horn are joined together by the *gray commissure*. The *posterior* horns are longer and narrower than the anterior horns and extend throughout the entire spinal cord. These are cells that mediate *pain and temperature*. There is another group of large neurons found at all levels of the spinal cord. They form the *dorsofunicular white matter*, which carries *tactile impulses* and *proprioception*. An important portion of the posterior horn is the *column of Clarke*. The column of Clarke extends to the entire length of the *thoracic* cord. Fibers from the *spinocerebellar tract* originate from this location. The posterior horns receive, through the *posterior lateral fissure*, the *sensory* part of the spinal nerve.

On the other hand, the *anterior* horns have projections that exit and form the *motor* root of the spinal nerve. The posterior and lateral columns come together to form the *anterolateral column*, also known as *intermedial horns*. The anterolateral column is a conglomeration of neurons that contains preganglionic fibers of the *sympathetic* division of the autonomic nervous system. At the level of the *sacral parasympathetic* nuclei is a conglomeration of *preganglionic* neurons, which integrates the *parasympathetic* column from **S2** to **S4**. From these cells (through the S2-4 ventral root) there are axons that extend to postganglionic neurons in the *pelvic* organs (which they innervate).

The *spinal cord commissure* is composed of two portions, a *white* portion and a *gray* portion. The *white commissure* is situated at the bottom of the anterior and medial sulcus. The *gray commissure* connects the two gray portions (anterior and posterior horns) at the midline. Along the entire length of the gray commissure is the *canal of the ependyma*, which (in the upper portion) is in wide communication with the *fourth ventricle*. The *posterior horns* are narrow at the level of the *cervical* cord, but the *white matter* is more abundant than the rest of the cord. In the *lumbar* region, the *gray matter* is more abundant.

Along the spinal cord there are *thirty-one* (**31**) pairs of *spinal nerves*. The first pair of spinal nerves exits from the spine between the atlas and the skull. From the level of **C2** to **C7**, the nerves leave the spinal canal *above* the vertebra with the corresponding number. There are **8** cervical nerves. The eighth spinal nerve leaves the spinal canal between the C7 and T1 vertebrae.

At the thoracic cord, there are **12** pairs of *thoracic* nerves. There are **5** pairs of *lumbar* nerves, **5** pairs of sacral nerves, and only **1** pair of coccygeal nerves. Each of the spinal nerves has a dorsal root ganglion (except for the first cervical and first coccygeal nerves). It should also be mentioned that there is NO *dorsal* root at the first cervical nerve.

The *white* substance of the spinal cord has primarily *longitudinal* fibers; however, there are also *transverse* as well as *oblique* fibers. There is a conglomeration of white matter cords. These cords, or fasciculi, each have a special function. The one in the *anterior lateral column* (at each side of the anterior medial sulcus) can be divided into the *direct pyramidal tract*. This tract is only found at the *cervical* cord (upper segment), and disappears completely at the upper dorsal region. There are decussated fibers that contain axons from the motor cortex. Fibers of this tract decussate along the course (across the midline) through the *anterior white commissure*. They have a triangular shape with the base located anteriorly. In the most anterior part of the anterolateral column, one can find the cross-pyramidal tract. As stated before, this tract is formed by a conglomeration of *descending* fibers which originates from the *motor* cortex of the frontal lobe (on the opposite side). They travel through the cerebral peduncle, pons, and medulla oblongata, and then decussate at the origin of the *first cervical nerve*. In the downward course, as they descend in the spinal cord, they decussate through the *anterior white commissure* to reach the cross-pyramidal tract of the opposite side. In the most anterolateral portion, an *ascending tract* lies in an angle formed by the cross-pyramidal tract and the direct cerebellar tract. It consists of centripetal *ascending* fibers, which originate from cells located at the base of the posterior horn cell. They quickly decussate to the opposite side at the anterior commissure and extend all the way up to the medulla oblongata, pons, and cerebellum. This tract is known as the *bundle of Gowers*. Another bundle of fibers is the *direct spinocerebellar tract*, which is peripherally located behind the bundle of Gowers and lateral to the cross-pyramidal tract. The direct spinocerebellar tract begins at the level of the *upper lumbar* region, and as it ascends, it thickens. It passes through the *inferior cerebellar peduncle*. Most of its fibers originate at the column of Clarke (at the gray matter, medial portion of the posterior horn cells).

Very near the point of entrance of the *posterior root* into the spinal cord, there is a tract known as the *tract of Lissauer*. The tract of Lissauer is *lateral* to the entrance of the previously described posterior nerve root. In fact, it is formed by many of the *ascending* fibers of the *posterior* root. These fibers run a short upward course to enter the posterior horn. In the *posterior column* of the spinal cord there are two tracts, the *tract of Goll* and the *tract of Burdach*.

The *tract of Goll* (which is more medially located) increases in size as it ascends to the medulla oblongata and reaches the *nucleus gracilis*. It contains *long* fibers that originate from the posterior root of the spinal nerve.

The *column of Burdach* is formed by *short ascending* fibers, also derived from the posterior root of the spinal nerve. This tract enters the gray matter of the posterior horn to form (medially) the *column of Clarke*. Other fibers run a medial course. These are the fibers that can be traced all the way up to the medulla oblongata. At the level of the cervical cord, as well as the most upper portion of the thoracic cord, attached to the most inner part of the column of Burdach, are a group of *descending* fibers known as the *descending virgular tract*, or *tract of Schultze*. These fibers are the descending portion of the dorsal nerve root.

From the *substantia gelatinosa* (which is a conglomeration of cells located at the apex of the dorsal horn) and from the neuron of the column of Clarke, originates a tract known as the *spinothalamic tract*. The spinothalamic tract carries *pain and temperature* impulses. The majority of these fibers swing ventrally and medially to *cross* to the opposite side of the spinal cord through the *ventral white commissure* to *ascend* forming the *lateral spinothalamic tract*, in the ventolateral portion of the white matter of the spinal cord. The fibers that carry pain and temperature are organized from the *lateral* portion to the *midline* in the following fashion. The sacral fibers occupy the most superficial portion, followed by the lumbar, thoracic, and cervical fibers. Therefore, the fibers that carry pain and temperature are the most *medial* ones. As mentioned before, the *lateral spinothalamic tract* ends at the most *posterior* and *medial* portion of the *thalamus*.

The *spinotectal tract* is another bundle that accompanies the lateral spinothalamic tract. This tract originates from the *substantia gelatinosa* of the posterior column and also decussates at the *anterior white commissure*, and in its outward course, runs *medial* to the lateral spinothalamic tract. As it ascends in the brain stem, it turns dorsally to the spinothalamic tract until it reaches the *superior colliculus*, where it is distributed.

The *ventral spinothalamic tract* contains axons with cellular bodies that are located in the dorsal root ganglia (DRG). As they enter the spinal cord, these axons bifurcate to the long ascending and short descending fibers to end at the *dorsal funicular nucleus* of gray cells (eight segments above the point of entrance into the spinal cord). This tract carries *tactile* function. The ventral spinothalamic tract ascends into the brain stem to terminate in the *ventroposterolateral nucleus* (VPL) of the *thalamus*.

In the pons, the spinal cord joins the *medial lemniscus* until its termination in the *thalamus*.

From the *cervical* segment of the spinal cord (from the most dorsal and medial cells of the posterior horn) arises the *spino-olivary tract*. These axons promptly decussate in the *ventral white commissure*, rather superficially, and *ascend* to the cervical cord and brain stem to enter the *inferior olivary nucleus*.

Besides the *pyramidal tract* and the *tectospinal tract*, there are other *descending* tracts. For example, we have the *rubrospinal tract*. The **rubro**spinal tract originates at the *red nucleus* to decussate immediately at the midline, and then runs lateral and posterior in the brain stem to enter the spinal cord. The majority of the fibers of the rubrospinal tract end at the *reticular formation* of the *brain stem*. The small portion that enters the spinal cord distributes itself into the *ventral* horn cells of the spinal cord.

The *uncinate fasciculus* (soon after they originate from the *nucleus fastigi* of the *cerebellum*) follows the inferior cerebellar peduncle, and enters the *ventral* funiculus of the spinal cord, to terminate at the *ventral* horn cells.

The *vestibular system* has a number of connections with the spinal cord and extends from the vestibular nuclei forming two important tracts, the *medial vestibulospinal tract* and the *ventrolateral vestibular spinal tract*.

The *medial* vestibulospinal tract originates at the *inferior vestibular* nucleus of the *medulla oblongata* to immediately decussate. They occupy the midline, from which some of the fibers decussate and descend into the cervical cord (just medial to the tectospinal tract), to end at the *motor* neurons of the *anterior horn cells*—which control the movement of the *neck* and *upper* extremities.

The *ventrolateral* vestibulospinal tract originates at the lateral vestibular nucleus and descends (without decussating) in the most *lateral* portion of the medulla oblongata and the spinal cord at many levels. They distribute in the *motor* neurons that control the movement of the *lower* extremities.

The *reticulospinal tract*, which takes origin in the cells of the reticular formation of the pons and medulla, descends to the medial reticulospinal pathway and in the *sympathetic* portion of the autonomic system of the spinal cord.

MYOTOMES AND DERMATOMES

As the central nervous system develops, *myotomes* migrate to form the *muscular* system of the adult. The *dermatomes* carry the innervation to a definitive *segmental pattern* to the skin surface of the arms, trunk, and legs. In the embryological studies of Haymaker, it has been shown that the buds of the embryo grow out to form the extremities. This growth carries the nerve impulses of the cervical segment for the innervation in the upper extremities from C4 to T2, and in the lower extremities from L2 to S3. Therefore, there will be a definitive *segmental* pattern of *muscle innervation*—which is very similar to the skin innervation. As the motor nerve root develops and grows out into the somites (that subsequently will form the different muscles of the adult), they carry specific innervation for a muscle. As the spinal cord subsequently dilutes in the *cauda equina*, the nerve from the cauda equina forms complex lumbar and lumbosacral plexuses (which somewhat complicate the understanding of the dermatome pattern).

The roots of the spinal nerve are attached to the spinal cord.

1. The *anterior* nerve roots are an outward continuation of the axons of the multipolar anterior horn cells.

2. The *posterior* nerve roots enter the spinal cord as two bundles; one ventral and one lateral. The *ventral* fibers form the column of Burdach, and the *lateral*, as they enter, are divided into *medial* and *external fascicles*. They enter the posterior horn through the *substantia genitanosa*. The most external bundle is *Lissauer's tract*.
3. All the posterior root fibers divide into ascending and descending branches as soon as they enter the spinal cord. They give abundant collaterals that form synopsis in the posterior horn cells, and another forms synopsis at the cells of the column of Clarke. The ascending fibers integrate with the *fasciculus cuneatus* and *gracilis* or *column of Goll* and *Burdach*, and then synopsis occurs at the cells of the nucleus cuneatus and gracilis at the *medulla oblongata*.

The spinal cord is a very important *conductor*, which carries motor and sensorial impulses. However, the spinal cord is also the center of many reflexes, some assure the muscle tone and the trophism. It also controls the sphincter, genital functions, and several sympathetic functions (like sweating and vasomotor responses). The spinal cord is the center for the cutaneous and deep tendon reflexes. The spinal cord is nourished by the anterior spinal artery branch of the vertebral arteries, the two posterior spinal arteries (also originated from the vertebral arteries), and the spinal artery branches. The thirty-one pairs of spinal nerves flow through the intervertebral foramina. The anterior-medial spinal artery is located in the midline and has a serpentine course downward. The *upper* three cervical spinal arteries are derived from the *vertebral artery*; whereas, the *lower* cervical branch arises from the *ascending pharyngeal* (at the level of the first rib) from the cervical intercostal trunk, and branches of the subclavian. The other branches for the spinal cord originate from the thoracic and lumbar aorta (from the common iliac and lateral sacral artery). Branches of the ascending cervical and the cervical intercostal trunk, (when they reach the medulla oblongata), are divided into ascending and descending branches, and anastomose among themselves. There is, at the point of anastomosis between the *anterior spinal artery*, a "watershed area" that is relatively vulnerable to ischemic changes.

Medullary arteries reach the spinal cord, and the largest branch of this artery (*artery of Adamkiewicz*) takes origin at **L2** on the *left* side. This artery follows an upward course along the L2 nerve root. Once it joins the anterior medial spinal artery, it turns downward to feed the *conus medullaris*. The intrinsic arteries of the spinal cord have two groups, *centrifugal* and *centripetal arteries*.

The *centrifugal* system has about 280 arteries that leave the medial anterior spinal artery at right angles. They run along the anterior and medial sulcus, then split right and left to give the blood supply to the central gray matter of the spinal cord. These branches are a conglomeration of capillaries that feeds: the anterior horns, the gray commissure, as well as the anterior two-thirds of the entire spinal cord. Branches from the pial plexus—which are branches of the intercostal—also directly enter the spinal cord, and do not anastomose with the capillary network. The blood supply from the column of Clarke is received from the *anterior spinal arteries*, but also from branches of the *posterior spinal arteries*. These posterior spinal arteries enter the white matter and posterior horn creating a dense plexus in the horn itself. The intramedullary arteries are *terminal* arteries (meaning that they do NOT anastomose).

There are several lesions that affect the spinal cord. For example, trauma, arthritis, herniated disk, or a tumor.

Trauma to the cervical cord can produce quadriplegia, respiratory difficulties, and changes in the blood pressure and sensory level. *Breathing* difficulties occur in lesions affecting the *phrenic nerve* nucleus from **C3** to **C5**, which may lead to paralysis of the *diaphragm*.

At height level, between C1 and C2, after *cordotomies* for the treatment of intractable *pain*, the patient may have paradoxical paralysis of *respiration*. This means that the patient will be able to breathe while awake and ceases to breathe while asleep—possibly due to damage to the *reticulospinal tract*. In trauma of the spinal cord (cervical and upper thoracic), the patient may have damage to the *intermedial lateral column* and present with splanchnic vasoparalysis—creating *arterial hypotension*. There is a relative increase of the vascular tree for the same amount of blood creating a relative hypovolemia. The lesion will also present with sympathetic paralysis, leaving a full command of the vagal nerve with subsequent *bradycardia*. The patient may not sweat below the lesion so the skin may appear hyperemic and warm. Sweating may occur in the face and eyelids only. At times, the same patient may present with *Horner's syndrome*, with *myosis* in the affected side of the spinal cord. This particular myosis should always be remembered because *most of the traumas to the spinal cord are accompanied by head injury*. So, the examiner may speak of unequal pupils and may pay more attention to the *midriatic* pupil—looking for intracranial pathology—when in reality, the abnormal pupil is the *small one* and indicates a *sympathetic dysfunction*.

In trauma to the spinal cord, *central cord syndrome* may occur as a consequence of *ischemia* to the cord. This is due to trauma of the anterior spinal artery (Schnider's syndrome). The patient may present with paralysis of the arm and still may be able to walk. It should be remembered that the cortical fibers (when they decussate at the anterior commissure) are more *centrally* located.

Lesions that involve the *lower* spinal segment affect the innervation and function of the *urinary* and *anal sphincter*, as well as the function of the urinary bladder.

The *urinary bladder* derives its nerve supply from two main sources, the *parasympathetic*, which carries volitional control, and the *sympathetic* nerve, which originates in the intermedial lateral column of the lower thoracic cord from **T10** to **L1**. They travel through the anterior root and extend to the paravertebral sympathetic ganglia. The postsynoptic sympathetic fibers descend along the aorta to form a bifurcation at the presacral nerve or hypogastric plexus. Their branches reach (through the hypogastric nerve) the urinary bladder. The parasympathetic supply of the bladder also arises from the intermedial lateral segment of the spinal cord, mainly from **S2, 3** and **4**. They form the *pudendal* nerve and the *pelvic* nerve, to end in the intramural ganglia of the bladder after they intermingle with the sympathetic fibers.

It should be remembered that *cortical* control for the bladder *contraction*, through the upper motor neurons (UMN), is located at the paracentral lobe and establishes synopsis at the *conus medullaris* (at the S3 and S4 segments).

The *pudendal nerve*, which originates from **S3** and **S4**, supplies the *sphincter* of the *urethra* as well as the *perineal muscles*. From the urinary bladder, there are fibers that carry the *sensation of fullness* of the bladder. These fibers travel through the pelvic nerve and the hypogastric nerve. The hypogastric nerve carries the sensation from the trigone of the bladder. At the level of the *pons*, there is an autonomic center where the fibers reach the bladder to exercise a facility influence for the *contraction* of the bladder. When the hypothalamus was described, it was mentioned that this organ sends fibers that travel through the spinal cord along the pyramidal tract to reach the bladder; these fibers are also *sympathetic* in nature and also have a facility function over the bladder.

NEUROGENIC BLADDER

When there is impairment of the function of the bladder, one speaks of neurogenic bladder. There are five types.

1. The *non-inhibited bladder*, which may occur in vascular lesions that affect the *paracentral* lobe and also can be produced by *tumors* as well as *demyelinating* diseases.

At the level of the *conus medullaris* is the center for micturation, ejaculation, erection, and defecation. When there is a lesion of the spinal cord *above* the sacral levels S2-3, the patient will present

2. with n*eurogenic reflex bladder*. It is called neurogenic reflex bladder because the function of the bladder depends only on the short *spinal reflex*. Two weeks after the injury, there is some automatism in the micturation, and it may be accompanied when the patient is stimulated. Then, the *mass reflex* will take place, in which the patient's bladder empties, priapism occurs, and there is flexure contracture of the lower extremities.

If, on the other hand, there is a destruction of the *sacral* segment, then the bladder function depends solely on the *intramural plexus* of the bladder, which results

3. in *autonomic neurogenic bladder*. In this case, the bladder musculature tends to become *hypertrophied*.

When the patient has a disease that affects the *anterior* horn cells, as in the case of *poliomyelitis*, the sensation for micturation is perceived, but there will be NO bladder contraction, and the bladder becomes distended and flaccid constituting

4. *motor atonic neurogenic bladder*.

If, on the contrary, the *posterior* horn cells are affected, as in the case of *tabes dorsalis*, the patient will theoretically be able to control his or her bladder.

However, the patient does not feel the sensation for micturation and this results

5. in *sensory atonic neurogenic bladder.*

OTHER FUNCTIONS

The *rectal sphincter* can also be disturbed in lesions that occur *above* the *sacral* level. *Voluntary* control of the *anal sphincter* is lost. If the lesion, on the other hand, occurs *at* the sacral cord, the patient will have fecal *incontinence* (where in the former there is fecal *retention*).

The ability to have an *erection* is due to engorgement of the *corpus cavernosum* and a good control that avoids the escape of *venous* blood. Impairment of the *sympathetic vasoconstriction* center at the level of **T12, L1-2** will also impair the erection. The impulses from this center travel through the *superior hypogastric plexus.* The contraction of the *seminal vesicle* also depends on the sympathetic system. Therefore, in cases of *pelvic sympathectomy,* ejaculation does NOT occur. As previously mentioned, in a lesion of the lower cervical and thoracic cord, engorgement of the corpus cavernosum take place due to the fact that the vasoconstrictor fibers that innervate the corpus cavernosum are not functional, and the patient will present with a *lasting erection* known as *priapism.*

In lesions of the *cauda equina,* many nerves as well as many sensory modalities can be affected. A *painful anesthesia* may occur. Spontaneous *pain* does occur in the affected nerve; however, when the nerve is explored, the patient is NOT capable of perceiving the noxious stimuli. Segmental groups of muscles may occur. Usually this syndrome is due to *trauma,* but it can also occur in a tumor or with a large *herniated nucleus pulposus.* In long-standing paralysis, ulcer of the skin can occur and a neurodystrophic phenomenon is observed (in which *osteolysis* can occur, with profound *decalcification* of the phalanges and the metatarsal bones).

In *acute traumatic paraplegia,* following anatomic or physiologic *transection* of the *spinal cord, complete* paralysis of the muscles *below* the lesion take place. In addition, there are sensory deficits for all sensory modalities *to the level of the lesion.* This is known as *spinal shock.* Spinal shock lasts approximately six weeks, and when the patient begins to improve, the activity of the *flexor* muscles appears before that of the extensor muscles.

If the transection of the spinal cord is *incomplete* and only *half* of the cord is affected, *hemiplegia* or *hemiparesis* will occur in the *affected* side as a result of the damage to the *pyramidal tract.* In the same side of the paralysis, there will be *loss of touch.* In the *contralateral* side, there will be *loss of the superficial sensibility for pain and temperature*; this is known as the *Brown-Sequard syndrome.*

Disturbances in *motility* that result from damage to the *pyramidal tract* can create marked *spasticity* of the muscles. There is great resistance to passive movement, which is more pronounced at the beginning of passive movement. But soon, the muscles give way and result in an *open jack-knife phenomenon.* In cases of paralysis, the deep tendon reflexes (DTR) are exaggerated and muscle clonus may occur—most commonly observed at the ankle and patella. The Babinski's signs or reflexes are *positive*. This means that after stimulation of the plantar portion of the foot, the big toe is *extended* (instead of curling downward as seen in a normal response). This reflex can also be explored by rubbing the skin at the *tibial* region, known as *Oppenheim's reflex*, or by scratching the *outer lateral* portion of the foot, known *Schaffer's reflex*. It can also be elicited with a percussion hammer tapping over the cuboid bone in the foot, a reflex known as the *Mendel-Bekhterev reflex*. Also, if one taps the toes, the patient will present an extended big toe, known as the *Rossolimo's sign*.

The *anterior horn cells* can be affected and may produce *complete* paralysis of the affected muscles. Rarely, one will find that all the muscles of the given extremity are involved. Among the most common entities that affect the anterior horn cells are *poliomyelitis* and the *muscular atrophy due to Werdnig-Hoffmann disease*. In most of the anterior horn cell diseases, vasomotor disorders are more frequently detected than in lesions affecting the pyramidal tract. Fasciculation is present in the muscles supplied by the corresponding affected anterior horn cells.

Anterior horn cell disease can occur simultaneously. The best example is the *amyotrophic lateral sclerosis* (ALS) or *Lou Gehrig's Disease*. It is a disease that affects, different muscles in a progressive fashion, which ends in *atrophy*, and there is absolute absence or lack of reaction in the muscles. This indicates that the *pyramidal tract* is affected at the *cerebral* origin of the *motor* cortex. In this entity, there is degeneration of the nucleus of the *cranial nerves* at the level of the medulla oblongata and/or the corticobulbar tract, affecting the muscles of the tongue, soft palate, pharynx, and larynx.

The *sensory* pathway of the spinal cord can also be affected. Among the diseases that affect the posterior root and the posterior sensory ganglion, is *syphilis*. In this particular instance, there will be a *loss of* pain, touch and temperature, and vibratory and position sense. The patient will present with *ataxia*, due to the marked impairment of *proprioception*. His or her gait will be worse if the eyes are closed or if he or she is in darkness, for the eyes are the crutches of the cabetic patient. These patients have *muscular hypotonia*. The *Romberg sign* is *positive* and the patient may have trophic ulcers in the plantar aspect of the foot. Their joints do NOT perceive pain, and profound *arthopathy* can occur with *deformities* of the affected joints; such a deformed joint is known as a *Charcot joint*. In many cases, one of the early presentations is *crisis of abdominal pain*, due primarily to the affected posterior ganglia.

If the lesion affects the *gray matter* or the *lateral spinothalamic tract*, the patient will have loss of *pain and temperature*, and can easily burn his or her finger and not be aware of it, particularly if the patient uses matches or is smoking. The patient has *thermoanesthesia* and *lack of pain perception*. A classical example is the *syringomyelia*. This abnormal cavity usually develops in the gray matter, in the most dorsal and lateral portion, and it can increase in size. The most commonly affected area is the *cervical* cord, and it may be seen in *Arnold-Chiari type II malformation*. Diseases that affect the *posterior column* may occur in combination with diseases that affect the *spinal ganglia*, such as the previously described *tabes dorsalis*.

Friedreich's ataxia is another example. This is a disorder in which there is loss of *deep* sensitivity, with involvement of the *posterior column* and the *pyramidal tract*. This disease usually begins at the early age of seven to eight years. Due to the fact that the posterior columns are effected, *proprioception* is lost. The patient may initially present with *areflexia*; however, with *bilateral* Babinski signs. *Tremor, nystagmus,* and bradylalia (slow speech and hesitant speech) have also been found in this disease.

There is a paralysis that usually begins in the *lower* extremities and subsequently involves the muscles of the trunk and upper extremities, and can also involve the *bulbar muscles*. This paralysis is *flaccid* and is due to a lesion of the *peripheral motor neuron*. It is usually *ascendent* and is known as *progressive ascendent paralysis of Landry*. This paralysis is due to an acute myelitis or polyradicular *myelitis*, which can occur after an upper respiratory infection. Fever, emesis, and flaccid paralysis are affected; however, the *sensory* system is *intact*. It can produce respiratory paralysis and death.

The herpes zoster virus also affects the posterior root and ganglia. It is characterized by severe *pain* in the affected nerve or root. There are also *vesicles* along the distribution of the affected dermatome. In many circumstances, once the acute episode passes, the patient may present with a severe neuralgia known as *herpetic neuralgia*.

Several diseases may affect many portions of the spinal cord, producing *motor* and *sensory* disturbances:

Multiple sclerosis is a disease in which the most typical thing is that nothing is typical! However, it produces diffuse *demyelinating plaques* over the central nervous system affecting both the *white* and the *gray* matter. The nerve fibers become demyelinated. It usually occurs before the age of forty years. It is a process that does NOT affect only the spinal cord. Due to the signs and symptoms that are present when the spinal cord is involved, it is important to describe the *spinal symptoms*. The patient can present with: paraplegia, intentional tremors, absence of the cutaneous abdominal reflexes, or a syndrome like Brown-Sequard. At the present time, it is easier to establish the diagnosis with the help of an MRI, somatosensory evoked potential, a study of the basic myelin protein in the cerebrospinal fluid, CSF protein electrophoresis, and with a visual and brain stem evoked response. The type of muscular atrophy that is observed in multiple sclerosis is due to lesions that affect the *peripheral* motor neurons. It occurs in the cell itself, usually affecting the muscles of the *thenar* and *hypothenar* regions. This gives the hand the appearance of a *monkey's* hand, known as a *simian* hand. This type of atrophy is known as *Aran-Duchenne atrophy or syndrome*. It should be pointed out that it can also be observed in *syringomyelia* and *amyotrophic lateral sclerosis*. It should be mentioned, for the sake of discussion only, independent of the spinal cord, there are *myopathic atrophies* that are hereditary, familial, and are never accompanied by sensory disturbance. These can produce many deformities, depending on the group of muscles that are affected. For example, it can occur in the facioscapular humeral muscles that involve the face, as well as the shoulder girdle muscles. Again, for clarification, it should be clearly understood that this last description does not correspond to spinal cord diseases.

A *familial* example of posterior and lateral column involvement is found in what is known as *subacute combined degeneration of the spinal cord*. This entity occurs as a complication of *pernicious anemia*, lack of Vitamin B12, as well as in patient's that have undergone *gastrectomy*. The dominant manifestations of this disease are: ataxia, loss of position and vibratory sense, spasticity, paraparesis or paraplegia, and areflexia. Usually, the patient presents with tingling in the upper extremities. The patient walks with a broad-based gait. The legs appear somewhat stiff with a lurching (bent) gait. The difficulty in walking is due to two problems. Primarily, it is due to involvement of the *corticospinal tract*, but also loss of proprioception due to *posterior column* involvement. This, by description, involves the posterior column and the pyramidal tract. However, it should be known that it also involves the *peripheral* nerves and the brain. The patient also presents with memory impairment and hallucinations.

The Cervical and Brachial Plexus 22

The spinal nerves are integrated by the union of the *anterior* and *posterior* roots. These roots originate through several rootlets from the spinal cord. The *anterior* one emerges from the anterior surface of the cord, and the *posterior* from the posterolateral sulcus. A small ganglion is located at the level of the foramen at each posterior root before they join the anterior root. The posterior root ganglion gives origin to fibers that enter the peripheral sensory fibers, and from the other pole, the cells of this ganglion enter the spinal cord. After the nerves emerge from the intervertebral foramina, they are divided into two branches. The *posterior* branch follows a backward course and supplies the skin and the posterior nuchal muscles (through medial and lateral branches). The *anterior* division of the spinal nerve is the one that forms the cervical and brachial plexus. The cervical plexus is formed by the anterior division of the upper cervical root. From this plexus, there are muscular branches that supply the anterior as well as the lateral flexor muscles of the head. The *rectus capitus lateralis muscle* flexes the neck laterally and also rotates the neck. The *longus colli muscle* flexes the upper cervical spine forward. The *rectus capiti anterior* is a short muscle that flexes the head anteriorly (see Figure 22-1).

There are also *cutaneous* branches that take origin in the superficial cervical plexus. The *greater occipital nerve* originates from **C2**. The *lesser* occipital nerve, which gives the *sensory* supply to the *back* of the head, also originates from **C2**. The *greater auricular nerve*, or auricularis magnus, takes origin from the second and third cervical roots, and this nerve supplies the lateral portion of the head and ear. The *transverse cervical*, which also originates from **C2** and **C3** (where there are branches that anastomose with cranial nerve XI and with the hypoglossal nerve), forms the *ansa hypoglossi* nerve that supplies the strap mucles of the neck.

The *supraclavicular nerve* originates from **C3** and **C4**. This nerve provides neural supply to the lateral and lower anterior portion of the neck, and to the clavicular region and right shoulder. From C2 and C3, there are branches that supply the *levator scapula.*

One important nerve that originates from the cervical plexus is the *phrenic nerve*, from **C3** and **C4**. It supplies innervation to the *diaphragm*.

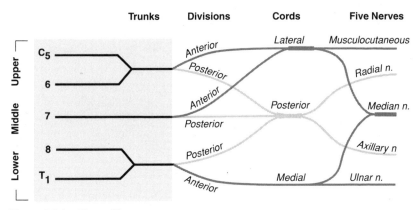

Figure 22-1 The axillary nerves.

Paralysis of this nerve produces *paralysis of the diaphragm*. Although the trapezius and sternocleidomastoid muscles receive their primary innervation from the spinal accessory (cranial nerve XI), they also receive innervation from the upper three cervical roots.

The brachial plexus is the result of the union of the anterior division of the fifth, sixth, seventh, and eighth cervical nerves and the first thoracic nerve.

The anterior division of the fifth and sixth cervical nerves join together to form the *superior primary trunk*. The anterior division of the seventh cervical nerve becomes enlarged and by itself forms the *middle primary trunk*. The anterior division of the first thoracic and eighth cervical nerves unites to form the *inferior, or lower primary, trunk*. Therefore, there are three primary trunks (upper, medial, and lower).

Each one of these trunks is divided into anterior and posterior branches. The anterior branches of the fifth, sixth, and seventh cervical nerves join together to form the superior and external cord, also known as the *outer cord*. This is where the musculocutanous and the outer roots of the median nerve originate. The anterior branches of the eighth cervical and first thoracic nerves join together to form the *inner cord* of the brachial plexus. From this cord is born the *ulnar nerve*, the *antebrachial cutaneous nerve*, the *medial brachial cutaneous nerve*, and the *inner root of the median nerve*. All the posterior branches join together to form the posterior cord of the median nerve, giving origin to the *posterior cord* of the brachial plexus. This gives rise to the *radial* and the *axillary nerves*. The upper part of the brachial plexus is located superior to the subclavian artery. To identify the lower portion of the brachial plexus, one should look behind the subclavian artery.

The brachial plexus has several branches.

The *long thoracic nerve* (originating from **C5, C6,** and **C7**), directly from the anterior root itself, supplies the *serratus anterior muscle*. Lesions of this nerve produce *winging of the scapula*.

The *suprascapular nerve* originates at the fifth and sixth cervical root and supplies the *supraspinatus* and *infraspinatus muscles*. The two muscles maintain the head of the humerus at the glenoid cavity.

The *anterior thoracic nerves* are two in number, and arise from the lateral cord of the brachial plexus, specifically from **C6, C7,** and **C8**. These nerves supply the *pectoralis major* and *minor muscles*.

The *dorsal scapular nerve* originates at the fifth cervical nerve and provides the nerve supply to the *rhomboid muscle*.

The *subscapular nerve* originates from **C6-7** of the posterior cord and specifically supplies the *teres majors* and the *subscapular muscles*.

The *thoracodorsalis nerve* originates from the **C6** and **C7** posterior cord and gives the nerve supply to the *lattismus dorsi*.

The *brachialis cutaneous medialis*, as well as the *medial brachiocutaneous*, originate at **C8** and **T1** to give the *cutaneous* nerve supply of the arm and upper forearm.

The *circumflex nerve*, also known as the *axillary nerve*, originates from the fifth and sixth cervical nerve and provides the nerve supply to the *teres minor* as well as the *deltoid muscles*.

The *musculocanateous nerve* originates at the lateral cord of the brachial plexus and provides the nerve supply to the *brachialis, choracobrachialis*, and *biceps muscles*.

The *radial nerve* also originates at the posterior cord from fifth, sixth and seventh cervical roots. The radial nerve provides neural supply to the triceps, anconeus, brachioradialis, the extensor carpi radialis longus, and brevis, as well as the supinator muscles. The extensor digitorum communis, indicis propius, and digiti quinti propius, as well as the longus pollicis, also receive their nerve supply from the *radial nerve*. The main function of these muscles is to *extend* the *fingers* and the *wrist*. The muscle extensor carpi ulnaris, the extensor policis brevis, and abductor pollicis longus also receive nerve supply from the radial nerve.

The radial nerve has *cutaneous* branches. The posterior brachialis cutaneous supplies the dorsum of the arm. The dorsal antebrachial cutaneus nerve supplies the skin over the dorsum of the hand and fingers (except half of the dorsal portion of the ring finger and the fifth digit). This nerve is very commonly paralyzed, particularly, with fracture of the humerus and shoulder, and/or by the use of crutches. There is a common type of paralysis that occurs when the arm is allowed to hang for hours over the back seat of the chair, this is known as the *"Saturday night palsy of the alcoholic."*

The *median nerve*, as mentioned before, has two heads or roots of origin, one from the inner cord and another from the outer cord. It supplies all of the flexor muscles except the flexor carpi ulnaris and the inner half of the flexor digitorum profundus. These *flexor* muscles are: the flexor carpi radialis, the palmaris longus, the pronator teres, the flexor digitorum sublimis, the flexor digitorum profundus, flexor pollicis longus, and the pronator quadratus. In the hand, it supplies the abductor pollicis brevis, the flexor pollicis brevis, and the opponens pollicis. It also supplies the second lumbricales muscle. Lesions of this nerve produce marked atrophy of the thenar eminence, and the flexor of the interphalangeal joints is absent.

The *sensory* abnormality found in the hand involves the palm of the hand, radial side, and the palmar surface of the thumb, index, middle, and ring fingers. In the dorsal aspect of the hand, the sensory supply of the medial nerve corresponds to the skin over the distal phalanges of the thumb, index, middle and the radial half of the ring fingers.

One common lesion of the *median* nerve occurs at the level of the wrist, within the *carpal canal*. Compression of the *median nerve* at this level produces the *carpal tunnel syndrome*. The patient can complain of pain of the wrist, radiating to the first three digits and superiorly to the anterior aspect of the forearm. The patient also presents with hypoalgesia over the hand distribution of the median nerve, and numbness in the first three digits that usually awakens the patient at night. The patient has to shake the hand to obtain relief. The *Tinnel sign* is positive at the anterior aspect of the wrist (pain and an electric shock-like feeling that occurs when the median nerve is tapped in the anterior aspect of the wrist). Atrophy of the thenar area and/or weakness of this muscle may be found. Flexing the wrist, in chronic cases, elicits *pain* over the median nerve distribution—this represents the *Phalen's sign*. The most common cause of the carpal tunnel syndrome is due to *inflammation of the synovia* of the *flexor muscles* at the carpal tunnel. It is also found in cases of: amyloidosis, hypothyroidism, occupations that require typing or excessive writing, diabetes mellitus, and arthritis.

The *ulnar nerve* arises from the inner cord of the brachial plexus and it supplies the flexor carpi ulnaris, the flexor digitorum profundus, the muscle of the hypothenar region, the abductor digiti quinti, the flexor digiti brevis, and the opponens digiti minimi. It also gives nerve supply to the abductor pollicis and to all the interosseous muscles of the hand and the two most medial lumbrical muscles. This nerve supplies the most medial portion of the flexor pollicis brevis and palmaris brevis. There is a *cutaneous* branch, which originates at the level of the wrist and supplies the dorsal aspect of the fifth digit and the ulnar half of the ring finger. Paralysis of the ulnar nerve produces a *claw hand* in which the metacarpophalangeal joints are hyperextended, but the interphalangeal joints of the last two fingers are semiflexed.

It is important to mention that in the course of the ulnar nerve, behind the medial epicondyle, the nerve may become entrapped. Fracture or trauma at this level produces paralysis that is rather slow and delayed, and is known as *tardy ulnar nerve palsy*. A lesion of the *median* nerve *and* the *ulnar* nerve can produce *causalgia*—which is accompanied by trophic changes of the skin. The skin becomes cyanotic and dry, and the nails can become very tough. The phenomenon of causalgia occurs more frequently with an incomplete lesion of the median nerve than any other nerve. On occasion, there occurs a syndrome known as *shoulder-hand syndrome*. This causalgia is accompanied by pain in the shoulder, stiffness of the shoulder, and swelling of the hand. The etiology of this syndrome is not clear. Some authors believe that it can be due to vasoconstriction of the peripheral arteries and pressure over the nerve root. These patients also present with purple discoloration of the fingers. The etiology is not clearly understood. At times, the skin may appear shiny and thin, accompanied by osteoporosis of the bones of the hands.

LESIONS OF THE BRACHIAL PLEXUS

These lesions may occur at birth, and also may be post-traumatic (as in a gunshot injury). The patient may present with *total* paralysis of the *arm*. At times, the brachial plexus may have an *incomplete* lesion, for example in the cervical ribs. This cervical rib arises from the *seventh cervical vertebra* and it can *compress* both the brachial plexus and the subclavian artery. This is also known as the *scalenous anticus muscle syndrome*. It usually produces numbness and coldness in the skin of the hypothenar region. Occasionally, we can observe cyanosis and dystrophic phenomenon of the skin.

Lesions of the brachial plexus can be divided into upper brachial plexus, lower brachial plexus, and lesions of the secondary cord.

Upper brachial plexus paralysis is known as *Erb-Duchenne palsy*. It is produced by damage to the fifth and sixth cervical roots, and will present with paralysis and atrophy of the deltoid, biceps, brachialis, long supinator, and supinator brevis muscles. The most common cause is from *traction injuries*. Usually, it does NOT present with sensory deficits. One common cause of a traction injury occurs during *delivery* (when the infant is pulled out).

Lesions of the *lower* brachial plexus result from damage to the first thoracic and eighth cervical roots. Of course it will involve all the muscles innervated by the ulnar nerve and the inner root of the median nerve. There will be paralysis of the flexors of the forearms. The hand appears like a *claw hand*. It can also occur during breech delivery.

Lesions of the *secondary cord* are NOT as common as the lesion of the *root*. If the outer cord is the one that sustains the injury, all the muscles supplied by the musculocutaneous and the outer root of the median nerve will be either *paralyzed* or *atrophic*. If the trauma or lesion is at the posterior cord, the obvious injury will occur in the muscles supplied by the radial and axillary nerves.

Finally, a lesion of the *inner cord* produces damage in the muscles supplied by the *ulnar* and *median* nerves.

The Lumbosacral Plexus 23

The *lumbar* nerves, of which there are five for each side, appear between the first and second lumbar vertebrae and extend to the upper part of the sacrum, at its base.

The roots of this lumbar nerve are the largest, and they are very closely attached to the lower end of the spinal cord. They are divided into a posterior and an anterior division.

The *posterior* division is divided into internal and external branches. The *external* branch supplies the intertransverse muscles of the spine and the erector spinae. The *internal* branch provides the nerve supply for the interspinalis muscle and the multifidus muscle.

The *anterior* division of the lumbar nerve, at its origin, communicates with the lumbar ganglion of the sympathetic by a slender filament. This nerve passes outward behind the psoas muscle. The *anterior* divisions of the *fourth* upper nerves anastomose between themselves to form the *lumbar plexus*. The anterior division of the *fifth* lumbar joins with branches of the fourth nerve, in front of the base of the sacrum to join the anterior division of the first sacral nerve. This forms the lumbosacral trunk, which will be studied together with the sacral plexus. Therefore, the lumbar plexus is formed by the anterior primary division of the very first three lumbar nerves and a major part of the fourth lumbar nerve. The plexus is situated in the psoas muscle itself in front of the transverse process of the lumbar vetebrae. The first lumbar nerve receives branches from the last thoracic nerve, T12. From this area, a large branch originates and is divided immediately into two nerves, the *ilioinguinal* and the *iliohypogastric*. It sends down a communicating branch that joins to a branch of the second lumbar to form the *genitocrural nerve*.

The second, the third, and the fourth lumbar vertebrae are divided into anterior and posterior branches. The anterior division of the second nerve is subdivided into anterior and posterior branches. The *anterior* division of this second nerve joins one branch (of the first nerve) to form the *genitocrural* or *genitofemoral nerve*.

The other branches of the second nerve join a branch of the third lumbar to form the *lateral cutaneous nerve* of the thigh. A branch of L2, L3, and L4 join together and form the *femoral nerve*.

Other descendent branches of L2 join with descending branches of L3 and L4 to form the *obturator nerve*.

The iliohypogastric nerve, which arises from T12-L1, extends laterally and obliquely in front of the quadratus lumborum to reach the iliac crest. The iliohypogastric nerve (at the level of the iliac crest) is divided into two branches, the *iliac* and *hypogastric*. The iliohypogastric nerve supplies the skin of the buttocks and the hypogastric region.

The *ilioinguinal* supplies the sensitivity for the medial part of the thigh, root of the penis, and upper part of the scrotum and/or labia majora.

The *genitofemoral* nerve supplies the cremesteric muscle and the labia majora in the female.

The *lateral femoral nerve*, which originates from the second and the third lumbar nerves, gives the nerve supply to the anterior and lateral portion of the thigh to the knee level. Sometimes, trauma to this nerve can produce a syndrome known as *neuralgia paresthetica*. In this syndrome, there is burning pain over the distribution of the nerve, accompanied by an objective sensory loss in the lateral part of the thigh. The most common cause is *compression* of the nerve as it passes through the inguinal fascia. The *femoral nerve* arises from the ventral division of the second, third, and fourth lumbar nerves and its sensory branch supplies the middle part of the anterior thigh through the cutaneous branches. Its largest portion is the *saphenous nerve*—which distributes in the medial portion of the leg to the ankle level. The muscular branch of this nerve provides innervation for the quadriceps femoris, the pectineus and sartorius muscles. The iliopsoas muscle is also suppled by this nerve.

The *obturator nerve* originates from the anterior division of the second, third, and fourth lumbar nerves, and supplies the skin innervation for the medial and lower portion of the upper thigh. The obturator nerve gives the nerve supply for the obturator externus, and abductor longus, brevis, and magnus.

The *superior gluteal nerve* arises from the fourth and fifth lumbar nerves as well as the first sacral nerve, and provides the nerve supply to the gluteus medius and minimus. From the fifth lumbar nerve and first and second sacral nerves, originates the *inferior gluteal nerve*, which supplies the gluteus maximus. The fifth lumbar branch and the first sacral nerve originate the nerves for the gemini, quadratus femoris, quadratus lumborum, piriformis, and obturator internus. As mentioned before, the fifth *lumbar* nerve joins the first *sacral* to form the *lumbosacral cord*.

The *sacral plexus* is formed by the *lumbosacral cord* and the *anterior* division of the three upper sacral nerves. The *posterior* divisions of the sacral nerve are rather small and are divided in upper internal and external branches to supply the multifidus spinae. The sacral plexus has a triangular shape with the base that corresponds with the exit point of the nerve at the sacrosciatic foramen. It rests over the *piriform muscle* and anteriorly, is covered by the *pelvic fascia*.

The *sciatic nerve* is the *largest* of all the nerves in a human. It arises from the branches of the fourth and fifth lumbar, and first, second, and third sacral nerves. It provides the nerve supply to the biceps femoris, semimembranous, semitendinous, and adductor magnus. This nerve is divided into two large branches, the *tibial nerve* and *common peroneal nerve*.

The *tibial nerve*, also known as the **internal popliteal nerve**, gives the nerve supply to the gastrocnemius, the plantaris, the soleous, and the popliteal muscles—which are essentially the *flexion* of the foot. This nerve also supples the tibialis posticus, the flexor digitorum longus, and the flexor hallucis longus, which *flexes* the *toes* and moves the foot *outward*. The flexor digitorum, the flexor hallucis brevis, the adbuctor hallucis brevis, abductor minimi, the flexor minimi brevis, as well as the interosseous and lumbrical of the foot are supplied by the *medial* and *lateral plantar* division of the tibial nerve.

The tibial nerve also gives cutaneous branches, such as the *medial sural nerve*, which supplies the skin of the lower third of the leg. The *calcaneus medialis branch* supplies the heel and the medial side of the sole of the foot. This portion of the skin is also supplied by another branch known as the *medial plantar nerve*. The *lateral plantaris nerve* supplies the lateral portion of the sole of the foot.

The *common peroneal*, also known as the **external popliteal nerve**, provides the nerve supply to the skin of the anterolateral and posterior portion of the leg by a branch known as the lateral sural nerve. The common peroneal nerve bifurcates to form the *deep peroneal* (or anterior tibial nerve) and the *superficial peroneal nerve*. The anterior tibial nerve innervates the anterior tibialis muscles, the extensor digitorum longus and brevis, the extensor hallucis longus, and the peroneal tertius muscles. Those muscles *extend* and *evert* the foot and toes. The *superficial* peroneal nerve supplies the peroneus longus and brevis, and the sensory supply to the *lateral* portion of the *ankle*.

The sciatic nerve, when affected, can produce *severe pain* that radiates into the distribution of the nerve itself or the radicular component of this nerve. This pain is known as *sciatica*. This pain can be produced by a herniated lumbar disk that compresses the fourth and the fifth lumbar nerves as well as the first sacral nerve. The pain is aggravated by straining, lifting, coughing, sneezing and other such movements. The herniated disk, however, is not the only cause for sciatica; it can also be produced by a *tumor* like a neuroma, abscesses, fracture, or trauma. The herniated disk is more frequently found at the *lumbar* region. The disk is nothing else than a cushion between two vertebrae, with an outer and inner part. The *outer* part is known as *annulus*. The center is *gelatinous* in consistency and called the *nucleus pulposus*. When the annulus breaks, the jelly leaks out and compresses the nerve. This leakage of the nucleus pulposus is called a *herniated disk*. It can occur centrally or laterally. The most common area for herniated disk is between the *fourth* and *fifth lumbar vertebrae* and between the *fifth lumbar* and *first sacral nerve*. In the neck, the most common location for herniation is between the fifth and sixth cervical nerves and sixth and seventh cervical vertebrae. By and large, initially, the patient may develop a pain known as *lumbego*, that can either improve or progress. Compression of the different parts of the sciatic nerve, at roots L4, L5, and S1, produce a pain that is *radiated* to the buttocks and the posterolateral aspect of the thigh and calf. If the **L5** nerve is the one compressed, usually the pain in the *foot* is located at the *extensor* aspect of the foot. If the *first sacral root* is compressed, the pain and sensory changes occur in the *lateral* part of the *foot* and *last two toes*. Paresthesia, tingling, and numbness are characteristic of *discogenic syndrome*, whether the herniation is cervical or lumbar. The numbness and tingling has the distribution of the nerve root that has been compressed.

In **L5-S1** *disk herniation*, the *ankle reflex* is usually *absent*.

In herniation at the **L3-4** level, the *patellar reflex* is *absent* and the patient can present with weakness of the quadriceps and numbness that occupies the lower third of the thigh and the inner third of the tibial region.

In a herniated disk at **L4-5**, paralysis produces a condition that is known as *footdrop*. When the patient is examined, if one presses the neck by gently compressing the internal jugular vein, that will increase the intraspinal pressure and produce a sciatic pain. In cases of a herniated lumbar disk, *Naffziger's sign* is seen. If the person has the leg elevated (straight leg raising), it also produces pain on the affected side; this is the *Lasegue's sign*. When the disk is centrally located, the pain is usually bilateral, the diagnosis is usually made, by MRI, CAT scan, or myelography in addition to clinical evaluation.

As mentioned before, herniation of a disk can occur in the *thoracic cord* as well in the *cervical cord*. Cervical disk herniations are *less* common than *lumbar* disk herniations. The herniation can also be central or lateral. The pain that occurs usually begins at the *cervical* region. From there, depending on which nerve root is involved, it will either radiate to the shoulder, scapular region, or arm. Compression of the nerve root number **5** will radiate to the *shoulder* and may produce numbness of the skin over the deltoid areas, weakness of the deltoid muscles, and, on occasion, the biceps. If the compression occurs at the cervical root number **6**, the pain will extend from the shoulder to the biceps, along the brachioradialis, and sometimes to the thumb or index finger. The paresthesia is usually in the same area of pain distribution.

The *biceps reflex* can either be decreased or absent. As a way to remember: the word **biceps** has *six* letters, which means that the **sixth** cervical root is compressed. If the triceps muscle is the one involved, it should be remembered that the word **triceps** has *seven* letters, implying that the **seventh** cervical root is the compressed. When the seventh nerve root is compressed, the pain from the shoulder radiates to the inferior angle of the scapula, triceps area, extensor aspect of the forearm, and the dorsal aspect of the three middle fingers. The *triceps reflex* is absent and the triceps muscles can be either weak or paralyzed, as well as the extensor muscles of the wrist and finger. In both circumstances (herniation of cervical nerves six and seven), movement of the neck aggravates the pain. If the disk is very large, it may compress half of the spinal cord, producing a *Brown-Sequard syndrome*. With this, there will be *upper* extremity paralysis *and* ipsilateral paralysis of the *lower* extremity. In this particular case, the reflexes of the lower extremity may be exaggerated, accompanied by a Babinski sign.

Central disk herniation may produce quadriparesis or paraparesis, with exaggeration of the deep tendon reflexes. The diagnosis is easily established by ordering a MRI and/or myelography with a post-myelogram CAT scan.

Thoracic disk herniations are less common, and they tend to compress the spinal cord and an intercostal nerve that is involved by the herniated disk. The thoracic disk has a tendency to *calcify* rather quickly. Diagnosis is established by performing myelography and magnetic resonance imaging (MRI).

INTERCOSTAL NERVES

The intercostal nerves, or dorsal nerves, are *twelve* in number for each side. The first dorsal nerve appears between the first and second thoracic vertebrae, and the last one appears between the last thoracic and first lumbar vertebrae. The anterior and posterior roots join at the intervertebral foramina and as they exit, they are divided into dorsal and anterior intercostals. The *dorsal* divisions of this nerve are usually *small* and have a backward course. They are divided into internal and external branches.

The *internal* branch of the first six *upper* thoracic nerves run inward giving the nerve supply to the multifidus muscle and the semispinalis dorsi. Then, after they pierce the trapezius muscle and the rhomboid muscle, they become purely *cutaneous*. The *lower* six thoracic nerves are purely *muscular* and supply the multifidus spinae, and they do not provide cutaneous innervation.

The *external* branches run through the longissimus dorsi and iliocostal muscles, giving innervation to these muscles as they pass through. The lower five nerves, after they pierce the rhomboid and trapezius, supply the skin that covers the rib cage.

The *anterior* divisions of the thoracic nerve, also called *intercostal nerves*, are also *twelve* in total for each side. They are distributed to give a nerve supply to the intercostal muscles and to the muscles of the abdominal wall. These nerves, in contraposition with the cervical and lumbar nerves, do not form plexuses. Each one of these nerves is connected with the *sympathetic* ganglia by one or two filaments.

The *lower* six thoracic nerves provide the nerve supply to the muscles of the abdominal wall and also give cutaneous branches to the gluteal region. It should be remembered that the first thoracic nerves also join, in part, the lower cervical nerves in the formation of the brachial plexus.

The six *upper* thoracic nerves give nerve supply to the intercostal muscles. They are located between the external intercostal muscles and the pleura. They also give the nerve supply to the internal intercostal muscles. The second intercostal nerve is the largest in size and crosses the axilla to reach the inner portion of the arm, and provides nerve supply to the skin at the most upper and medial part of the arm. The *last* thoracic nerve is the *longest* of the thoracic nerves, and it provides anterior branches that innervate the lower intercostal muscles and the diaphragm. It runs between and in front of the quadratus lumborum, and runs between the internal oblique and fascia transversalis and the internal oblique to subsequently join the lumbar plexus—the first lumbar branch to form the *iliohypogastric nerve*.

The Autonomic Nervous System   24

The autonomic nervous system is an *in*voluntary portion of the nervous system that is divided into two portions.

1. The *craniosacral*, also called the *parasympathetic* system, is integrated by cranial nerves III, VII, IX, and X; and lower down by the second, third, and fourth sacral nerves.
2. The *thoracolumbar*, or *sympathetic* system, which occurs from the first to the twelfth thoracic nerve, and by the first and second lumbar nerves.

The thoracolumbar portion is also called *orthosympathetic*, and the craniosacral part is called *parasympathetic* (see Figure 24-1).

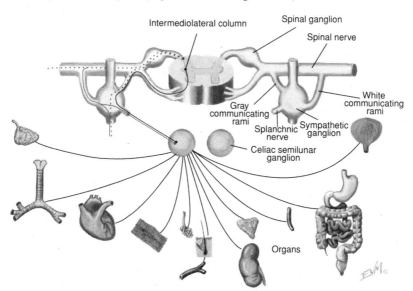

Figure 24-1 Sympathetic and parasympathetic innervation.

These nerves belong to the vegetative nervous system and give the blood supply to the wall of the blood vessels, the lungs, the heart, the bladder, and the glandular structures. Their innervation is concerned with the involuntary function of the body and with the complex coordination and modification of the organs to stress and environmental factors. This system establishes the complex balance between the internal and external activity of the human body. However, there is a very intimate relation with the voluntary function of the body.

THE SYMPATHETIC SYSTEM

The sympathetic system is **also called** the *thoracolumbar outflow*. It is the peripheral system of the autonomic system, which extends from the base of the skull to the coccyx. There is a pair of chains of nodular ganglia located in the anterolateral surface of the vertebral bodies. In the cervical region, there are only three and at the thoracic region, two of the thoracic ganglia are usually fused. They are connected with the internodal nerve trunk.

The paravertebral *chain ganglion receives fibers that are myelinated from the spinal nerve and are known as white communicating rami* (with the exception of the cervical and lower lumbar roots, which do not have white rami communication). The white rami carry visceral *a*fferent fibers and also axons from the intermedial lateral horns at the thoracic and lumbar regions. This intermedial lateral column has a lateral group of cells that are known as sympathetic superior and lateral nucleus. Some of these cells conglomerate more closely to the posterior horn forming the intermedial gray matter. From here, they send axons that leave the spinal nerve after the junction of the anterior and posterior root and through the white rami, communicating, they enter the paravertebral ganglia. It should be mentioned that each white ramus innervates more than one sympathetic ganglion. The cervical ganglia receives fibers from the upper and middle thoracic root.

The sympathetic chain is contained within ascending and descending myelinated and unmyelinated fibers, which enter and leave the ganglia at various levels. These fibers are both pre- and postganglionic. From each ganglion, a gray (unmyelinated) fiber has a retrograde course to enter, again, into the spinal nerve. This gray ramus distributes peripherally in the same area of the nerve that it has joined and brings vasomotor, pilomotor, and pseudomotor impulses. In addition, visceral fibers from the abdominal and thoracic viscera (that are innervated by the sympathetic chains) enter the central nervous system through the communicating rami and the posterior root of the spinal nerve. These afferent fibers do not actually originate at the sympathetic ganglia—instead, they pass through the ganglia. The cells of the paravertebral ganglia are large and spherical. Each ganglion cell has a single axon, which usually wraps around the cell; however, its dendrites are multiplex. The axons of the ganglion cells are unmyelinated fibers, and they pass anteriorly to join with the fibers of other ganglia to form a true trunk. An example is the splanchnic nerve, which subsequently forms complicated visceral plexuses. Among these plexuses are: the celiac, the mesentery, or hypogastric plexus, which provide the connection for central impulses. From these large plexuses, there are fibers that travel along the large blood vessels to form a network that will follow each of the branches that go to various abdominal or thoracic viscera. The relationship of the arteries with the nerve that travels with them is very intimate.

PARASYMPATHETIC BULBOSACRAL OUTFLOW

The sacral outflow originates at the gray substance of the second, third, and fourth sacral segments at the intermedial lateral column, forming the inferolateral sympathetic nucleus. From this nucleus, myelinated fibers leave the spinal cord through the anterior root and go directly to the urinary bladder, colon, genitalia, and uterus without entering the paravertebral ganglion chains, but they form synopsis with the intramural ganglion cells of the viscera that they innervate.

A great outflow of vegetative fibers originates from the nucleus of the vagus nerve at the *medulla oblongata*. The most anterolateral portion, known as nucleus ambiguous, has a special visceromotor function that is concerned with the striated muscles. Preganglionic fibers leave the medulla oblongata and through the vagus nerve enter the ganglion nodosum—to follow the entire course of the nerve to reach the abdominal and thoracic viscera. These fibers are directed to the heart, lungs, aorta, and abdominal viscera, and they end in intramural ganglia located in the wall of the viscera. From this intramural plexus, emerges short non-myelinated fibers, which supply the smooth muscles of the viscera that they belong to.

From the inferior salivatory nucleus of the medulla oblongata, there are fibers that leave the central nervous system via the glossopharyngeal nerve, to pass through the tympanic plexus and petrosal ganglion. From there, the fibers enter into the lesser superficial and petrosal nerve to the otic ganglia, and then through the auriculotemporal nerve to the parotid gland.

The superior salivatory nucleus is located in the reticular substance that is adjacent to the facial nerve, from where fibers leave the central nervous system to the nervus intermedius. Some of those fibers also leave the facial nerve and enter the chorda tympani to join the lingual nerve, to enter the submandibular glangion, and to distribute from there into the sublingual and submandibular glands.

Other fibers leave the facial nerve as the greater superficial petrosal nerve to end in the sphenopalatine ganglion. From this ganglion, originate fibers, which innervate the lacrimal glands as well as the glands from the mucosa of the nasopharynx.

At the cerebral peduncle, between the anterior portion of the two nuclei of cranial nerve III, there is an anterior and medial group of cells that constitute an anterior and medial nucleus, from which fibers go to the pupillary sphincter. These fibers leave the third nerve, in the orbital cavity, through its inferior division to join the *celiary ganglion*. From here, postganglionic fibers form the short celiary nerves that supply the smooth celiary muscles as well as the muscles of the iris. In general, this parasympathetic system liberates an adrenalin substance (NE) and therefore, the fibers are known as adrenergic fibers. Others liberate choline-like substance (Ach) and are known as cholinergic fibers. These substances, which mimic the adrenergic effect of sympathetic stimulation, are known as sympathomimetic; those that mimic the action of the *cholinergic effect of the parasympathetic nerve are known as parasympathomimetic.*

THE CERVICAL SYMPATHETIC

The cervical portion of the sympathetic trunk is located ventral to the transverse process of the vertebral bodies, and is located behind the common carotid artery. There are three ganglia.

1. The *superior cervical ganglion* is located in front of the transverse process of the second, third, and fourth cervical vertebrae and joined by the first fourth cervical nerves, the glosspharyngeal, the vagus, the hypoglossal nerve, and the pharyngeal plexus. From the superior cervical ganglion takes origin the superior cardiac nerve, which will join the cardiac plexus. It also gives origin to the carotid nerves, which wrap around the carotid artery to form an important sympathetic plexus that will follow each of the branches of the internal carotid artery.
2. The *middle cervical ganglion* supplies gray rami to the fifth and sixth cervical nerves. The middle cardiac nerve and fibers to the inferior thyroid artery arise from this ganglion.
3. The *inferior cervical ganglion* sends gray rami to the seventh and eighth cervical nerves, to the inferior cardiac nerve, and branches that form the plexus of the subclavian and vertebral arteries. Very often, the inferior cervical ganglion fuses with the first thoracic ganglion to form the stellate ganglion.

From the upper thoracic cord, and through the white rami, communicating, are sympathetic fibers that enter the sympathetic trunk to ascend into the cervical sympathetic trunk. Compression of the sympathetic ganglion gives origin to an ipsilateral Horner's syndrome, which has been described, and is characterized by miosis (small pupil), **en**ophthalmos (recessed eyeball), anhydrosis (lack of sweating), and ptosis (drooping eyelid).

The heart receives sympathetic fibers from each of the three cervical ganglia and all the thoracic ganglia from T1 to T7. This nerve, when stimulated, produces tachycardia (whereas the branch of the parasympathetic vagus nerve, produces bradycardia).

The lungs, *trachea*, and bronchi receive parasympathetic supply from a branch of the vagus nerve that ends within the intramural ganglia of the organ that it supplies. Stimulation of the sympathetic produces bronchial dilatation (whereas stimulation of the vagus produces bronchial constriction).

The nerve supply of the entire digestive system arises from the abdominal branch of the vagus nerve and from various segments of the thoracolumbar outfllow. The main sympathetic ganglion is the celiac ganglion. The epigastric, or solar, plexus is also known as the celiac plexus, and is located behind the stomach, in front of the aorta, and surrounds the celiac trunk of the aorta and extends as low as the pancreas. This plexus and the ganglia are interconnected and receive the small splanchnic nerves on both sides, as well as branches from the right vagus nerve. The solar plexus is composed of the two semilunar ganglia and are the largest ganglia in the body. They represent, indeed, an aggregration of the smaller ganglia. They also supply the pancreas and adrenal glands. The vagus nerve establishes connection with the ganglion cells and within the ganglion located at the wall of the abdominal viscera. Within the wall of the viscera are located the enteric plexuses that are responsible for the function of the bowels.

The sympathetic system innervates the smooth muscle fibers of the blood vessel walls. There is one special innervation for the carotid sinus. The carotid sinus is an important structure that maintains systemic arterial tension and the adequacy of cerebral circulation. It does respond to intravascular pressure. When the pressure increases within the sinus, the sinus responds by lowering the systemic blood pressure. Afferent impulses passing over the carotid sinus via a branch of the glossopharyngeal nerve reach the brain stem to produce a depressor effect. Stimulation of the carotid sinus produces a fall in the blood pressure and slows the heart rate. Patients that have a *hypersensitive carotid sinus may suffer from syncopal episodes.*

The *sweat glands* are innervated by the *thoracolumbar* outflow of the *sympathetic* nerve. At the medulla oblongata and hypothalamus are located sweat centers. From these originate pseudomotor fibers to the face that reach first the intermedial lateral column of the thoracic cord. They establish synopsis in the superior cervical ganglia, from which the fibers reach the face through the nerves that are located at the sheath of the external carotid artery.

The *urinary bladder* has two outer muscular layers that form the detrusor of the bladder. It has an inner longitudinal group of fibers situated at the level of the *trigone*. The *external* sphincter of the urinary bladder is composed of *striated* muscles fibers and it surrounds the urethra immediately distal to the prostate gland in men, and in the proximal portion of the urethra in women. The detrusor muscle, as well as the prostatic urethra, are supplied by the parasympathetic sacral nerve, which is called the *nerve erigentes*, which arises from the second and the third sacral roots. The fourth sacral root enters the inferior hypogastric plexus and gives the nerve supply to the plexus located within the bladder wall, from which postganglionic fibers supply the bladder musculature.

The sympathetic nerve to the bladder arises from the lower thoracic, first, second, third, and fourth lumbar segments by the white communicating rami to the paravertebral ganglia. They form synopsis at this ganglia and, through the mesenteric nerve, enter the *inferior mesenteric ganglia*. Then, they form fibers that produce contraction or retraction of the blood vessels of the penis, fibers to the seminal vesicle, and an important branch, the *nervus hypogastricus*, that joins the hypogastric plexus. From the hypogastric plexus also arise branches to the prostate gland and the gland of Littre of the urethra. The somatic innervation arises through the pudendal nerve from the third and fourth sacral roots to reach the muscles of the perinne. The spinal cord center of this double bladder innervation is located at the conus medullaris. The sensation of *bladder distention* is carried primarily by way of the pelvic nerve to the sacral cord and, from the ascendent afferent pathway located at the most posterior and superficial part of the lateral column, reaches the cortex of the paracentral lobe.

The *vegetative* or *autonomic involuntary nervous system* can be affected by a variety of conditions. There is a rare disorder manifested by redness and pain in the extremities. The pain can be of great intensity and is like a burning sensation. It can be precipitated by extreme cold. The feet are most commonly affected, and it usually is not symmetrical. This entity is known as *erythromelalgia*.

Peripheral vascular disturbances of the sympathetic nervous system produced by disturbance of the sympathetic system include those of spasms of the small arteries of the extremities. *Raynaud's disease usually occurs when the*re is exposure to extreme cold. It can be manifested also in the fingers, toes, or ears, and paresthesias occur in the anatomic part involved. Cyanosis can also present in the extremities that are involved. The attack is of variable duration, and at times, small vesicles are formed over the involved skin. These vesicles can be superficial or deeper, and they may form a dry gangrene. Sometimes, it is difficult to make a differential diagnosis from erythromelalgia.

Another disturbing vasomotor disorder is the *Raynaud's phenomenon*. This is a *symptom* and it can appear in many conditions (like *scleroderma* and/or *rheumatoid arthritis*). *Scleroderma* is characterized as a disorder in the *collagenous tissues* of the *skin*; although, it can also appear in the blood vessels that feed the central nervous system, the heart, and gastrointestinal tract.

Many patients may present with a condition characterized by excessive sweating of the hands and feet. The etiology is not clear, for which reason it is also essential *hyperhidrosis*.

A *progressive facial hemiatrophy, accompanied by Horner's syndrome*, may or may not be accompanied by *injuries* of the *sympathetic* plexus in the neck. This progressive disorder is known as *facial hemiatrophy*. The skin and tissues of the face, as well as the bone on the affected side of the face, can shrink and the hair becomes sparse (either in the beard, head, or eyebrow region). Many times it can be accompanied by atypical *facial pain*.

Brain Tumors   25

Brain tumors represent a very small percentage of all human tumors. Although it is not always possible to make the clinical diagnosis and localize the tumor, it is now possible—with the help of a MRI and CAT scan—to precisely locate a tumor. Intracranial tumors can be classified into gliomas, pituitary adenomas, meningiomas, acoustic neuromas, metastatic tumors, congenital tumors, blood vessel tumors, granulomas, sarcomas, and tumors of the choroid plexus.

GLIOMAS

To better recall the pathology of gliomas, keep in mind that the glial tissue has five types of cells: astrocytes, oligodendroglia, ependymal cells, Schwann cells, and microglia. If one adds the suffix "oma", we have astrocyt**oma**, oligodendroglioma, ependymoma, the Schwannoma, and microglioma.

Other gliomas found in the brain are: medulloblastoma, astroblastoma, spongoblastoma polare, pinealoma, ganglioneuromas, and neuroepithelioma.

The gliomas represent a group of tumors that originate in the glial matrix of the central nervous system. Eighty percent of gliomas usually fall into one of three categories: *astrocytomas*, *glioblastoma multiforme* and *medulloblastomas*.

ASTROCYTOMAS

Astrocytomas represent 37 percent to 38 percent of all gliomas. They usually occur in the *cerebral* hemisphere in *adults*, and in the *cerebellum* in the children. This tumor can either be solid or cystic. Although this tumor has similar histology, regardless if it is in the brain or in the cerebellum, the behavior and activity is somewhat different. *Astrocytomas* of the cerebellum are the most benign gliomas of the brain. These tumors, most of the time, can be cured with total resection of the tumor. A very high percentage of these cerebellar astrocytomas are cystic in nature. They can have a neural nodule in the wall of the cyst, within the cyst. They contain a very heavy xanthochromic fluid. The tumors are usually avascular and fairly well circumscribed. In the brain, however, astrocytomas are much less circumscribed, and in many instances, invade more than one lobe and sometimes, extending through the corpus callosum, they invade the opposite hemisphere.

Malignant *changes* can occur *within* this benign tumor. There are two types of astrocytes: the *fibrilar* and the *protoplasmatic* type. If the protoplasmatic cell becomes enlarged, as usually occurs in this type of tumor, it is called an astrocytoma or gemistocytoma. These do not confer malignancy or benign characteristics. If, in the fibrilar type of astrocyte, the fibrils become hypertrophied, they are known as the *fibers of Rosenthal*. Also, because this type of tumor, *fibrilar astrocytoma*, occurs in *children*, it is called *juvenile* astrocytoma.

It should be remembered that these fibers of Rosenthal are NOT pathognomonic of the fibrilar astrocytoma because it can also be found in some leukodystrophies. Some astrocytomas, in particular the protoplasmatic type, may undergo malignant changes and as mentioned before, they occur mainly in *adults*. The astrocytomas, in general, have system manifestations of increased intracranial pressure and specific manifestations, depending on which part of the brain is involved. The early manifestations may include: headache, vomiting, double vision, papilledema, and/or hydrocephalus if the tumor involves the regular pathway of the cerebrospinal fluid drainage. They also can produce seizures or paralysis, nystagmus, and ataxia, if located in the *cerebellum*. If a posterior astrocytoma involves the brain stem, the patient can present with different cranial nerve and sensory pathways (see *brain stem syndrome*). Astrocytomas are notorious for producing microscopic tumoral *calcification*.

OLIGODENDROGLIOMAS

The oligodendrogliomas represent 3.8 percent to 4 percent of all gliomas. This tumor presents with intratumoral *calcifications*, which are usually visualized in an x-ray. By and large, they are solid and occur mainly in the *cerebral* hemisphere of the *adult* between the ages of 34 to 45 years. This tumor can also occur in the thalamus or hypothalamus. In this tumor, mitotic figures are uncommon and the tumoral calcifications are usually perivascular. The clinical history of this patient is usually extremely long.

EPENDYMOMAS

The ependymomas represent approximately 3.7 percent of all the gliomas. They usually occur in *children*, and they are usually intraventricular. Although they can occur in any of the ventricles, they are more common in the *fourth* ventricle. Intratumoral calcifications are very common in the fourth ventricle. They are composed primarily of two cell types, *ependymal cells* or *ependymoblasts*. Both type of cells present as a rather small round or rod-shaped body within the cystoplasm. When this type of tumor occurs *above* the *foramen magnum*, it tends to be *malignant*; however, the same tumor with the same histology has a completely and totally different biologic behavior when it occurs in the spinal cord. The clinical manifestations are usually late due to the fact that they grow into the ventricular system.

Microgliomas

The microglia is a cell that has a *multipotentiality*, it behaves as "a lymphatic cell" in the brain. It may not be sufficient justification to classify a tumor of the microglia as *microgliomas* (keeping them unclassified). The microgliomas are, more commonly, found in patients that have AIDS and tumor growth in the brain.

Glioblastoma Multiforme

This tumor represents the approximately 30 percent of all the gliomas. It is a rapid-growing tumor of the cerebral hemisphere in the adult. *Glioblastoma multiforme* are usually located beneath the cerebral cortex, and infiltrate and destroy the brain tissue as they grow into the normal brain tissue. The margins of this tumor are rather indistinct, and it is literally impossible to establish a line of demarcation between the tumor and the normal brain. This tumor has the tendency to extend to the opposite hemisphere through the *corpus callosum*. At the vicinity, as well as at a considerable distance from the tumor, one can find marked *cerebral edema*. This tumor may have foci of hemorrhage, great *cellularity*, mitotic figures, necrosis, and endothelial proliferation in the blood vessels.

Multinucleated giant cells are usually found within this tumor. This tumor commonly manifests itself by *convulsions*, and according to the part of the brain that is invaded, may result in aphasia, hemiplegia, and personality changes. At the present time there is no satisfactory treatment. I have been using, for 23 years, after internal decompression, Decadron LA in the tumor bed followed by radiation. In many instances, I have obtained survival as long as 15 years. At present, the value of the intratumoral chemotherapeutic agent is questionable.

Medulloblastomas

These tumors represent 12 percent of all the gliomas. Medulloblastomas occur in the *cerebellar* hemisphere and in the first decade of life (usually in a male child). They also occur in the vermis of the cerebellum. It is believed that they arise from the external granular layer of the cerebellum. At times, they have a clear margin of demarcation. They tend to occlude the *fourth* ventricle, producing *hydrocephalus*. Many of the cases have presented as a *gastric outlet obstruction*, probably due to pressure over the nucleus of the *vagus nerve*. Medulloblastomas migrate along the subarachnoid space and can spread upward to the base of the brain or downward to the medulla, spinal cord, or cauda equina. Because patients affected by this type of tumor usually develop hydrocephalus, a *ventriculoperitoneal shunt* is part of the treatment. Unfortunately, metastases may be found in the peritoneal cavity.

In the past, medulloblastomas were considered as an undifferentiated developmental stage of the neuroepithelium (medulloblasts). They are typically composed of many packed cells with very little cytoplasm. In some cases, it is even difficult to demonstrate the presence of cytoplasm. These medulloblasts tend to differentiate in a variety of cell types; bipolar, unipolar, and astroblasts, since these cells are found within the tumor. One of the early presentations of these tumors is *projectile* vomiting. A patient may also present with *nystagmus* and a *staggering gait* (which may be confused as a result of weakness produced by the constant vomiting). Bilateral palsy of cranial nerve VI is common. In addition, other presentations include: dysmetria, dysdiadochokinesis, papilledema, muscular hypotonia, and the patients tend to fall backward. Many of the *sarcomas* of the *vermis* are believed to be due to desmoblastic changes of the medulloblastoma.

ASTROBLASTOMASA

These tumors represent 5 percent of all giomas and usually occur in the cerebral hemisphere of the adult in the fourth decade of life. Astroblastoma tumors are noncapsulated and they are very invasive. Microscopically, they are composed of an embryonic type of cell known as an astroblast.

SPONGIOBLASTOMAS

These tumors are also known as *spongioblastoma polaris*, and they represent approximately 4.5 percent of all gliomas. Spongioblastomas are found in the cerebral hemisphere, also in the brain stem and cerebellum. The optic chiasm and optic nerve are commonly involved, as well as the hypothalamus. When they occur in the pons, they show a typical appearance that is known as pseudohypertrophy of the pons. The fourth ventricle is usually NOT involved or shows only a very *late* involvement. Therefore, hydrocephalus is NOT common in this type of tumor. These tumors are by far the most common in the optic chiasm and brain stem. It should be remembered that gliomas of the optic nerve and optic chiasm are associated with neurofibromatosis multiplex.

PINEAL TUMORS

The tumors represent 2 percent of the gliomas and are mainly of two types, pineoblastomas and pinealomas. They can occur at any age, but usually occur in young males. The tumor of the pineal gland falls into four catagories: pinealomas, pineoblastomas, ganglioneuromas, and teratomas. These tumors extend forward into the third ventricle, or they can also compress the lamina quadrigemina producing hydrocephalus, and may also extend upward toward the splenium of the corpus callosum.

Pinealomas usually have two types of cells. Some have little or no cytoplasm and other cells have an epithelioid appearance.

Pineoblastomas have a tendency to seed themselves through the ventricular system and rarely, in the subarachnoid space. They may produce a Parinaud's syndrome (paralysis of the upward gaze and lack of convergence due to pressure upon the quadrigeminal plate). Due to the invasion of the dorsomedial and ventromedial nucleus of the hypothalamus, this patient can show *precocious puberty*.

The ganglioneuromas or gangliogliomas are extremely rare in the brain. They are usually found in the floor of the third ventricle, in the tuber cinereum. They have mature and immature cells, and clinical manifestations are primarily related to their location. On certain occasions, the patient can present with testicular and ovarian atrophy, and on other occasions, they present with hypersexuality. I had a patient whose wife divorced him because of his hypersexuality. This patient subsequently progressed to have visual hallucinations. He was "seeing" seven women who wanted to "rape" him. He offered to marry the most vociferous one. The remaining six women went in a riot against him. He started to shoot at the imaginary women, and he was taken to a mental institution. The tumor was found to be invading the *floor* of the *third ventricle*.

CHOROID PLEXUS PAPILLOMAS

These tumors represent less than 1 percent of all intracranial tumors. They are more commonly found in the *choroid plexus* of the fourth ventricle, but also can be found in the lateral ventricle. Like the ependymomas in the lateral ventricle, these papillomas occupy primarily the foramen of Monro, mainly on the left side. This creates *hydrocephalus*, primarily due to *excessive production* of cerebrospinal fluid.

COLLOID CYSTS

Colloid cysts may represent a remnant of the paraphysis, which originates at the junction of the ependyma with the tela choroidea at the roof of the third ventricle. This tumor can grow to a gigantic size, and tends to occlude the third ventricle or the foramen of Monro, producing hydrocephalus. One of the most common presentations is *severe headache*, usually preceded by syncopal episode when the patient flexes the head forward. The headache is thought to be due to acute hydrocephalus produced by the occlusion of the foramen of Monro. A sudden movement or shaking of the head presumably dislodges the cyst that was entrapping the circulation of the cerebrospinal fluid at the foramen of Monro. Although in many textbooks this seems to be the common presentation, I have, in my experience, operated on 70 cases of colloid cysts, and noted that only 2 cases had such a clinical presentation. In certain cases, the patient may present with seesaw eye movements, which is presumably due to stimulation (posteriorly) of the preoptic and paraventricular nuclei of the hypothalamus. This cyst contains a gelatinous material, which at times is somewhat hard. The wall of the cyst has a layer of cuboidal and ciliated cells. These are *benign* cysts and they are rather rare. Some authors have described them in the fourth ventricle as well as the lateral ventricle. In the past, the diagnosis was established with pneumoencephalography. Now, the diagnosis is easily made with a CAT scan or MRI of the brain.

INTRACRANIAL MESENCHYMAL TUMORS

This tumor of the connective tissue forms a large group that can not be incorporated in one full term.

Meningiomas, in particular, show a great predilection of occurring in *women*. Some of these meningiomas are classified by the *location*, rather than by the histology. When they occur in the sphenoid wing, they are classified as inner medial and outer third meningiomas. When they occur around the falx cerebri, they are called parasagittal meningiomas; and when they occur at the junction of the falx cerebri, falx cerebelli, and tentorium, they are called "quatre fleur meningiomas" or four-leaf meningiomas.

There are several types of meningiomas according to the histology.

1. *Mesenchymatous* meningioma, usually originate at the leptomeninges and can occur anywhere.
2. *Meningotheliomatous* meningiomas are the most common of all the meningiomas. They are usually very well encapsulated and present a rather slow growth, and are easily removable. They occur more commonly in the anterior third of the falx cerebri, in the olfactory groove, and sella turcica. They also occur along the lesser wing of the sphenoid.

These tumors can vary in size (either small or large). They have a firm and fleshy appearance. Very rarely, one finds mitotic figures in this tumor; however, one can find groups of organized tumor cells that are very frequently calcified, known as *psammoma bodies*. The meningotheliomatous meningiomas do not invade the brain tissue but they do invade the *dura mater* and sometimes, the bone. The so-called *en plaque* meningioma is the result of hyperostotic proliferation of the bone. Meningiomas of the sella turcica, by their location, tend to produce compression of the optic chiasm with *bitemporal hemianopsia*.

Those meningiomas arising from the olfactory groove tend to produce *anosmia* on the side of the tumor. They may also compress the optic nerve producing *blindness* and atrophy on the side of the compression with contralateral papilledema (Foster-Kennedy syndrome). Meningiomas that occur in the sphenoid wing may grow intraorbitally producing exophthalmos, and if they affect the cranial nerve going through the superior orbital fissure (cranial nerves III, IV, V, and VI), they may produce complete *ophthalmoplegia*.

Parasagittal meningiomas produce seizures, paraparesis, and urinary sphincter incontinence.

1. *Angioblastic meningiomas*, known as the tumor of Cushing and Eisenhardt, are rather rare. Although benign histologically, they can metastasize to the bones, lungs, and other areas like the breasts and the pancreas. These tumors grow rather rapidly and tend to recur even after an apparent total removal. These tumors are very vascular and they present mitotic figures. They have a tendency to occur along large venous channels and sinuses.
2. *Fibroblastic meningiomas*. These types of tumors can occur within the cerebral hemisphere itself, and are very common in *children*. The majority of the meningiomas that occur in the ventricular system are the *fibromatous* type. They also have a great tendency to produce seizures or paralysis.
3. *Lipomatous meningiomas*. The tumors may occur in any part of the central nervous system, but they are more commonly found on top of the *corpus callosum*.

4. *Sarcomas*. These are rare tumors that occur in the *leptomeninges* of the adult brain. They invade the meninges. They also occur in the *cerebellum*. Sarcomas metastasize through the cerebrospinal fluid to other parts of the central nervous system.

Melanomas of the brain can be either primary or metastatic.

LIPOMAS

The lipomas are rather rare tumors and usually occur on top of the corpus callosum, but also in the lamina quadrigemina and pituitary gland. They are commonly found at the spinal cord.

SCHWANNOMAS

The most representative of this type of tumor is the *acoustic neurinoma*. Acoustic neurinomas arise from the sheath of the *vestibular* portion of the acoustic nerve. They may involve the entire cerebellopontine angle and are highly associated with neurofibromatosis multiplex (von Recklinghausen's disease). They can begin within the internal auditory canal and expand to involve the cerebellopontine angle. They may produce unilateral deafness and paralysis of cranial nerve VII. According to the size, the tumor may also involve cranial nerve V. When this occurs, those patients quickly lose the corneal reflex. If the tumor compresses the brain stem, it may also produce paralysis or paresis, ataxia, and increased intracranial pressure. The symptoms are usually of long duration. The patient may, for years, hear noises within the head, intermittently or continuously; this is known as *tinnitus*. The compression of the cerebellum may produce a staggering gait and clumsiness of the hand on the tumor side and papilledema and/or paralysis of the lateral rectus muscle of the eye. Microscopically, these tumors are composed of elongated spindle-shaped cells forming rows known as *pallisades*. The diagnosis (besides performing audiology tests) is made by means of MRI or CAT scan.

TUMORS OF THE PITUITARY GLAND

The pituitary gland is a rather small but powerful gland located in the *sella turcica*. It is separated from the rest of the intracranial cavity by the *diaphragm sella*. There is an opening through this dural diaphragm, which the pituitary stalk exits to join the hypothalamus. This gland receives its blood supply from branches of the intracavernous portion of the carotid arteries. The sympathetic nerve supplies, arrives to the glands accompanying the arteries that give the blood supply. The sympathetic nerve distributes into the epithelial portion of the pituitary.

The pituitary gland (hypophysis) has two distinct portions: the *posterior* portion or *neuroepiphysis*, which is the result of an evagination of the floor of the third ventricle; and the *anterior* portion or epithelial part, which derives from an evagination of the Rathke's pouch. The *neuroepiphysis* contains a special part of the glial cells known as pituicytes, and is composed of nerve fibers that descend from the tuber cinereum and to the pituitary stalk, and also from fibers of the supraoptic and periventricular nucleus of the hypothalamus.

The *neuroepiphysis*, or *posterior* portion, secretes antidiuretic hormone (ADH) or vasopressin, and oxytocin. The antidiuretic hormone promotes the *absorption* of water in the distal portion of the renal tubuli. This hormone *inhibits* the thirst mechanism. When ADH is not secreted, the patient may urinate a lot and will drink a lot of water; this phenomenon is known as *polyuria* and *polydipsia*. ADH is produced in the *hypothalamus* and is transported through the axons of the tuber cinereum and pituitary stalk to the pituicytes, where it is stored. The discharge of the hormones is regulated by the serum osmolarity.

The epithelial neuroepiphysis has three portions: the pars tuberalis, pars distalis or *anterior* lobe. It is at this portion of the gland that *craniopharyngiomas* arise. The pars intermedia controls the melanocytes. The *anterior* lobe of the pituitary gland produces hormones that have great importance in the endocrine function of the human body. These hormones are:

- Growth hormone (GH)
- Thyroid stimulating hormone (TSH)
- Adrenocorticotropic hormone (ACTH)
- Follicle-stimulating hormone (FSH)
- Luteinizing hormone (LH)
- Prolactin

The regulation for the production of these hormones is through the hypothalamus.

The epithelial portion of the pituitary gland has three different type of cells.

- Acidophilic
- Eosinophilic
- Basophilic

The eosinophilic cells are concerned with growth. A tumor of these cells can produce *gigantism* or *acromegaly*. The basophilic cells produce the hormone that controls sexual function. The function of the chromophobe cells is not well known.

When the pituitary gland is hyperfunctioning, it is known as *hyperpituitarism*, and it may produce *gigantism* in the child and *acromegaly* in the adult. The acromegaly results in an increase in the size of the jaw, prognathism, macroglosia, big feet and hands, and large frontal sinuses. The nose widens and enlarges. The patient usually has a "barrel" chest. Diabetes mellitus is commonly associated with acromegaly and is usually a type of diabetes that is insulin-resistant. These patients may present with enlargement of the liver, intestine, stomach, and heart. *Acromegaly* usually occurs in tumors that involve the *eosinophilic* cells. In an x-ray, there is an enlargement of the sella turcica. The patient may present with a decrease in libido or menstrual irregularities. *Hypopituitarism* results from a lesion that involves the *anterior* pituitary gland. The tumor that most commonly produces the hypopituitarism is the *chromophobe adenoma*.

In some cases, when hemorrhage occurs immediately postpartum, necrosis of the anterior pituitary gland can occur producing a panhypopituitarism known as *Sheehan's syndrome*. These patients become lethargic and have loss of libido, tire very easily, and lose the axillary and pubic hair. As a conseqence of hypopituitarism, the patient can also present with hypo*thyroidism*. The electrolyte balance is usually maintained during hypopituitarism.

Pituitary dwarfism occurs when the pituitary function is destroyed during adolescence. The tumor of the hypophysis can extend beyond the sella turcica and may compress the optic chiasm. The microscopic appearance of this tumor can either be *chromophobe* type, *basophilic* type, and/or *eosinophilic* type.

1. *Chromophobe* adenomas are the most common pituitary tumors. They may create pressure over the optic chiasm producing bitemporal hemianopsia or blindness. The *prolactin*-secreting tumors are the most common tumors of the pituitary gland in women.
2. The *basophilic* adenomas are usually rather small and they may produce elevation of the ACTH. They are associated with Cushing's *syndrome*. This syndrome is characterized by moon facies, buffalo hump of the neck, reddish skin strias, amenorrhea, arterial hypertension, hyperglycemia, and sometimes, osteoporosis. This syndrome is likely due to adrenocortical hyperfunction, whereas Cushing's *disease* is produced by hyperfunction of the basophilic cells of the pituitary gland. If the patient has Cushing's *disease* and, by error, the adrenal glands are removed, the pituitary gland may grow uncontrollably. The skin color becomes similar to a person that has a permanent suntan. There is hyperpigmentation of the lip, nipples, and scrotum or vulva; this is known as *Nelson's syndrome*. Some of these pituitary tumors can grow very large and they may invade the hypothalamus and temporal lobe. The diagnosis of this pituitary tumor is assisted by means of MRI or CAT scan.

The *craniopharyngiomas* represent about 4 percent of all intracranial tumors. They occur primarily in children and are derived from the embryonic squamous epithelium of the hypophysis. These tumors may either be fleshy or cystic. They are notorious for presenting with multiple intratumoral calcifications. Craniopharyngiomas have been given several names, Rathke's pouch tumor, adamantinomas, and some other authors call it cholesteatoma. Usually, it is a very *large* cystic mass and it has a rather small solid portion. This tumor grows out of the sella turcica and compresses the optic chiasm, producing blindness. Furthermore, it can grow near the hypothalamus and produce hydrocephalus. The cystic content of this tumor usually has a yellow viscous, yellow-to-dark fluid, and contains a large amount of cholesterol crystals. This tumor usually appears before the age of 15 years and these children usually complain of headaches, vomiting, and failed vision. In many instances, these children are obese and do not grow or develop normally. Adipose genital dystrophy (Frolich's syndrome) is associated with a tumor that involves both the area of the pituitary as well as the hypothalamus. Due to the intimate attachment of this tumor to the hypothalamus, it presents a rather difficult problem for total removal. A partial removal of the tumor is always followed by recurrence.

The treatment of pituitary adenomas should be geared towards the *preservation of vision*. Sometimes, the treatment should be a combination of surgery and radiation therapy. Hormonal substitution, particularly thyroid and/or gonadal, is of great help as a substitute therapy. Certain tumors, like the prolactin-secreting tumor, can be amenable to treatment with bromocriptine mesylate 2.5 mg three times a day. This drug is known commercially as Parlodel. Also used are Dostinex tablets. These tablets contain cabergoline, a dopamine receptor agonist. The secretion of the pituitary gland is mainly under hypothalamic inhibitory control, likely exerted through the release of dopamine by the tubuloinfundibular neurons. Cabergoline is a long-acting *dopamine*-receptor *agonist* that *decreases* the serum *prolactin* level. Usually, four weeks after treatment, the prolactin level becomes normal. The recommended dosage is 0.25 mg twice a week.

The surgical approach for a pituitary tumor can be trans-sphenoidal, if it is a microadenoma and transfrontal, beneath the frontal lobe, if the tumor is extends above the sella turcica. If we are looking at the pituitary gland through the trans-sphenoidal approach, for the sake of memorizing, let us remember that the breasts are anterior and one for each side; therefore, looking from below, the prolactin-secreting tumors should be anterior and lateral. Since the adrenal glands are posterior and lateral, one for each side, therefore, ACTH producing tumors are also located posterior and lateral as the adrenal gland, analogizing the pituitary gland with the human body. This means, repeating again, that these ACTH-producing pituitary tumors are located *posteriorly* and *laterally* at the *anterior* lobe of the pituitary gland. Since in the human the thyroid is in the midline, again, if you look at the pituitary gland from below, the tumor that has low TSH-secreting hormone is located *anterior* and in the *midline* of the pituitary gland. If the tumors produce growth hormones, they are located behind the TSH-secreting area.

TUMORS OF CONGENITAL ORIGIN

These tumors are rather uncommon and arise from embryonic rest. Congenital tumors include:

1. Chordomas. Chordomas arise from a remnant of the notochord, usually in the basilar canal. They usually grow beneath the brain stem, pushing it backward, and may also extend down through the bone into the nasopharynx. This tumor can be found (although it is rare) in the *cervical* region, but can also be found in the *sacrum*. The microscopic appearance is that of large cells with vacuolated cytoplasm. The clinical manifestation depends primarily on the compression of the brain stem and the adjacent cranial nerves. Large chordomas may

even extend to the pituitary gland and optic chiasm. On x-ray, destruction of the base of the skull is common.
2. Epidermoids. Epidermoid tumors are also known as pearly tumors or cholesteatomas. They can occur within the two tables of the skull bone, diploe, and usually produce erosion of the two tables of the skull. They may produce compression of the dura and underlying brain. Cholesteatomas are the result of desquamated epithelium that arise secondary to inflammation of the ear and are not a true neoplasm.
3. Dermoids and teratomas. They are very rare and tend to be microscopically identical to those that occur in other parts of the body. These contain sebaceous material, hair, and squamous epithelium. Some of these tumors contain nerve fibers, sebaceous glands, muscle fibers, cartilage, and adipose tissue. These tumors occur mainly along the midline and often occur within the pineal gland.

TUMORS OF THE BLOOD VESSELS

These tumors represent 1.8 percent of all intracranial tumors. They are divided into two main groups, the arteriovenous malformation and the true vascular neoplasm. Among the malformation tumors, we have: (1) cavernous hemangioma, (2) capillary telangiectasias, and (3) arteriovenous malformation. These are usually found in some of the neuroectodermal diseases. They are an integral part of Sturge-Weber disease. This syndrome is characterized by hemangiomas of the face in the distribution of the fifth nerve, in one or all the branches of this nerve distribution. This syndrome can also be accompanied by seizures and/or mental retardation. On x-ray, calcification runs parallel, in a railroad track appearance. Telangiectasias can also be part of a neuroectodermal disease known as *ataxia telangiectasias*. The children affected with this illness can have telangiectasias of the conjunctiva and cerebellum. They may look normal otherwise, but as they progress in age, they have a severe cystic degeneration of the cerebellum, and they usually do not live more than six years.

The vascular neoplasms, like the hemangioblastomas, are also associated with another neuroectodermal disease known as von Hippel-Lindau disease. This disease results in retinal hemangiomas, liver angiomas, and a cerebellar tumor known as hemangioblastoma. The retinal arteries are rather large and tortuous, and end in a nodular angioma of the retina. They can produce degeneration of the retina. Sometimes, these angiomas may occur in the kidneys and adrenal glands. They tend to be hereditary. The cerebellar symptoms, at times, are minimal and at other times, are florid. They are produced by a tumor that is usually cystic and usually has a mural nodule. Removal of the tumor and nodule result in a cure.

Other vascular tumors include angioblastic meningiomas (see previous description).

Arteriovenous malformations are the most common vascular anomalies found intracranially. They are mostly found in the brain and are formed by a conglomeration of abnormal enlarged veins and arteries. They are usually cortically located. When they bleed, they tend to produce *seizures* (while ruptured intracranial aneurysms do NOT produce seizures). These large abnormal venous channels may draw blood from many arteries of the brain. There are arteriovenous fistulas that produce a bruit audible during cranial auscultation. These lesions can produce a seizure disorder, intractable headache, and repeated bleeding. Other clinical manifestations depend on the location of the malformation.

Spontaneous subarachnoid bleeding, as a consequence of the malformation, is less common than intracranial aneurysm. With diagnostic tools, CAT scan, MRI, MRA, and cerebral arteriogram, the diagnosis can be easily made. Also, with new surgical techniques like intravascular navigation, it is possible to occlude the malformation without opening the head. With the help of the surgical microscope, removal of this vascular malformation can also be accomplished. A cerebral aneurysm may be so large that it can be called an *aneurysmal* tumor.

METASTATIC TUMORS OF THE BRAIN

These tumors can arise from any part of the human body. Primary tumor of the lung will likely metastasize to the brain, and usually the metastasis occurs at the junction of the white matter with the cerebral cortex. Other tumors, like the ones that occur in the breast, kidneys, intestines, and thyroid can also metastasize to the brain. They can be solitary or multiple. *Metastasis* is the most common of all *cerebellar* tumors in adults. The *hemangioblastoma* is the most common primary *benign* tumor of the *cerebellar* hemisphere in the adult. This tumor can also metastasize to the leptomeninges, producing carcinomatosis of the leptomeninges. When the metastasis occurs in the meninges, the patient can present with hydrocephalus, headaches, vomiting, stiff neck, seizure, and meningeal signs. Carcinoma of the nasopharynx can grow inward at the base of the skull, in the intracranial cavity. These tumors are notorious for producing cerebral edema, sometimes disproportional to the size of the tumor. Initially, this tumor may appear as a nodule but very often as the tumor grows, it becomes necrotic in the center and may present, in the CAT scan or MRI, with a great similarity to a brain abscess.

Tumors of the eye, like the retinoblastoma, neuroepithelioma, or melanoblastoma can commonly metastasize to the brain.

TUMORS OF THE SKULL

This group of tumors is contained primarily in the bone itself, for example benign osteomas, but sometimes, they can extend beyond into the inner table and compress the dura and subjacent brain tissue. Hand-Schuller-Christian disease is a disorder that consists of a defect in the membranous bone. It can produce exophthalmos or diabetes insipidus if it invades the hypothalamus. The primary problem consists of a multifocal proliferation of the reticular tissue. These granulomas occur in the skull and contain a yellow material filled with lipid droplets. The course of this disease is rather chronic and the prognosis is uncertain.

Eosinophilic granulomas resemble Hand-Schuller-Christian disease and produce *osteolytic* lesions in the skull. At times, they are palpable through the skull. Both of these entities may respond to roentgen (radiation) therapy.

INFECTIONS OF THE BRAIN

1. *Brain abscess*: A pyogenic abscess of the brain occurs more often when there is previous damage of the brain—like an embolism. It likely will occur with: a pulmonary infection or an infection of the mastoids and sinuses. It can also occur in any part of the brain. If the abscess does not invade the meninges and/or the ventricular system, it usually does not present with fever. But, when it becomes advanced and there is seepage of the purulent material, it can produce *leukocytosis*. Once more, it should be mentioned that fever and leukocytosis may be *absent*. The clinical manifestion is increased intracranial pressure, and depends very much on what part of the brain it is located in. The diagnosis is rather easy with CAT scan and MRI, once the early stage of cerebritis has progressed to a full pyogenic cavity. It should be remembered that *headache*, by and large, has no localizing value except in brain abscess and subdural or epidural empyemas. The patient usually points where the headache is.

   Certain infections of the meninges, like the one produced by *fungi*, can also be manifested by increased intracranial pressure and may produce an abscess cavity. The torula histolytica, also called *cryptococcus neoformans*, is the single most common fungus that invades the meninges at the base of the skull. The diagnosis can be obtained with cisternal puncture to identify fungi. The coccidial granulomas are rather common in the cerebellar hemisphere and very commonly are associated with pulmonary infection.

2. *Infestation*: *Tapeworm infestation* of the brain can produce a large cystic cavity that can be intraparenchymal and intraventricular and in the meninges. The infestation with the taenia solium in the human is due to the ingestion of eggs present in the human feces. In the past, it was thought that it was rare in the United States, however, due to the large amount of immigrants into the United States and tourism by Americans to the underdeveloped countries, it has become more common. Eating infested salads or infected pork meat can also develop into a full-blown picture of *neurocysticercosis*. If *intraventricular* neurocysticercus does not respond to medication, it should be surgically removed. In contrast, the *intraparenchymal* as well as the *meningeal* neurocysticerosis are very sensitive to *praziquantel*, also known by the commercial name of Biltricide. Biltricide is a *trematodicide* that comes in tablets of 600 mg. It induces a rapid contraction of the larva due to the fact that the cell membrane is permeable to this

drug. I have treated about four thousand cases of neurocysticercosis and use 50 mg per kilo of body weight divided into three doses for twelve days. When the neurocysticercus occurs in the fourth ventricle, it can produce hydrocephalus. In Latin patients with seizures and positional paroxsymal nystagmus, one should rule out a neurocysticercus of the fourth ventricle.

3. Tuberculomas: Tuberculous infection of the CNS occurs in two forms, in the form of meningitis and tuberculomas. Meningitis produces headache, stupor, stiff neck, Kernig's sign, and characteristic cerebrospinal fluid findings of high protein, low sugar, and low chloride. Tuberculous meningitis can produce hydrocephalus, nausea, and vomiting due to the increased intracranial pressure. The tuberculomas can occur in the brain as well as in the cerebellum and/or in the brain stem. When they occur in the cerebellum, it is usually in children. The clinical manifestation depends very much on the location of the tuberculoma. With modern treatment of tuberculosis, tuberculomas can be totally irradicated.

4. Syphilis: Syphilis involving the brain and/or the meninges can produce granulomas known as *gumma*. Syphilis is the great simulator. It is due to invasion by *treponema pallidum*. Sometimes it is difficult to obtain a positive history of syphilis. A patient, in the period of invasion, usually complains of headaches and occasionally, a stiff neck. Spinal tap reveals increased cell count and elevated globulin. The VDRL test has contributed for the diagnosis of syphilis. Sometimes, however, there is false positive reaction. The study, the treponemal immobilization test, also called TPI, is specific for syphilis. The initial changes from syphilis are found in the blood vessels, which are infiltrated with lymphocytes and plasma cells, and can also produce thickening of the arachnoids. What is really occurring is *syphilitic arteritis* that can occlude the lumen of the arteries. Invasion of the cerebral parenchyma is one of the most serious of all form of the neurosyphilis. It is known as *general paresis* and is a late form of syphilis that usually occur many years after the primary lesion; the leptomeninges can appear thickened and cloudy; the patient may present with hydrocephalus and brain atrophy. These patients present with psychotic and neurologic problems. The general paresis can present with dementia, agitation, depression, and defect in judgment. Visual hallucinations are rather common and are accompanied by emotional changes; the patients may become obstinate and stubborn. The most common type of paresis is the dementing form, where there is profound degeneration of moral and ethical standards. They become apathetic and forgetful; sometimes this can be preceded by delirium of grandeur and quickly can go to a state of agitation and reach a maniacal state. Argyll-Robertson pupils are commonly found in general paresis. A syphilis-infected patient may present with slurred speech, tremors, and deterioration of the handwriting. If it affects and occludes the blood vessels, the patient can have a stroke, which is known *Lissauer's paralysis*, which is commonly associated with

seizures. It should also be remembered that syphilis can also produce *tabes dorsalis*. This usually occurs around twenty years after the primary infection. There is a profound deterioration of the posterior root, ganglion, and posterior column. In tabes dorsalis, crisis of pain can be an early symptom. The patient complains of rheumatic-like or muscular pain; common during cold weather. This pain is like lightening pain. Sometimes this pain has a belt or girdle characteristic and it is usually very sharp, and probably is due to irritation of the posterior root or posterior ganglia. At times, the pain can be in the rectum, larynx, and/or in the stomach. The gastric crisis is the most frequent. At times, the patient has facial neuralgias and at other times, has hypoalgesia over the bridge of the nose and cheeks in a butterfly-shaped area. The pain sensitivity is the one most affected—both superficial and deep pain. In these patients, one can squeeze the testicle and he would have no pain. If the examiner is exploring the pain, it is not unusual that upon applying a pinprick, it may take one or two minutes before that patient acknowledges the perception of the pinprick; it is indeed a *delayed* pain sensation. The vibratory and perception senses are equally and profoundly affected. Due to the damage of the posterior column, *ataxia* is a characteristic sign, more so if the patient closes the eyes or if it is dark. It can be said that the eyes are the crutches of the tabetic patient, and he or she falls when he or she closes the eyes (Romberg's sign). These patients also develop hypotonia and joint hyperextensibility. The deep tendon reflexes are usually lost or decreased. There is a special type of arthropathy known as Charcot joint, which can also be found in syringomyelia and peripheral neuropathy secondary to diabetes. It usually occurs as the consequence of loss of pain sensibility. Cranial nerves III, IV, and VI are usually involved—although optic atrophy can also occur. Gumma of the brain is a very rare manifestation of syphilis; it may simulate a glioma. It is usually located at the brain surface and it can elicit or produce a seizure. Focal symptoms depend on the gumma location.

SPINAL CORD TUMORS

The tumors of the spinal cord are subdivided into three catagories: (1) extradural, (2) intradural but extramedullary, and (3) intramedullary. Intraspinal tumors represent only one-sixth of the intracranial tumors. Intradural extramedullary tumors have the greatest incidence and represent 54 percent of the total percentage. Extradural tumors represent 21 percent of the total percentage, and intramedullary tumors constitute 25 percent of the total percentage. Note: the intradural but extramedullary tumors are primarily benign meningiomas (45%) and neurofibromas (55%). They occur primarily in females, in the first decade of life, and are usually located in the thoracic spine. They are not common in children, but not rare.

The second most common location of the intradural but extramedullary of the neurofibromas and meningiomas are (in order): cervical, lumbar, and sacral regions. The intramedullary tumors occur more often in the lumbar, thoracic, and cervical regions.

INTRADURAL EXTRAMEDULLARY TUMORS

These tumors arise from the sheath of the nerve as well as from the meninges, and they are primarily neuronomas and meningiomas. They are commonly associated with *neurofibromatosis* multiplex; sometimes they can extend through the intervertebral foramen to the intercostal nerve, forming a dumbbell type of tumor. Both of these tumors are encapsulated, and they are easy to separate from the spinal cord. The meningiomas are usually more vascularized and present intratumoral calcium deposits. In the removal of these meningiomas, it is always advised to remove the portion of the dura to which the tumor is attached.

EXTRADURAL TUMOR

The most common tumor of the extradural space is *metastasis*. In fact, any time that a patient who has a primary tumor any place in the body presents with evidence of *spinal cord compression*, one should think immediately of an *epidural metastasis*. These tumors tend to invade the vertebral bodies, the pedicles, and sometimes, the lamina. They can erode the lamina and invade the adjacent paraspinal muscles. The most common tumors that metastasize to the epidural space are those which originate in the lungs, breasts, colon, and thyroid; also, in the epidural space, can be found the lymphomas, Hodgkin's, and multiple myelomas. The overall prognosis of these tumors is rather poor and, if the compression has advanced to the point that the only part of the sensory modalities that persists is the touch, the possibilities of regaining normal motor function from paraplegia are almost nonexistent. If some motor activity is recovered, it would never be significant enough for the patient to ambulate.

INTRAMEDULLARY TUMORS

Ependymomas are the most predominant of all the glial tumors of the spinal cord, followed by *astrocytomas, oligodendrogliomas,* and, rarely, a *glioblastoma multiforme*. Occasionally, *medulloblastomas* can be found from intramedullary implantation from medulloblastomas of the cerebellum.

Other intramedullary tumors are hemangioblastomas, hemangiomas, and hemangioendothelioma.

All the tumors of the spinal cord and/or the spinal canal have a similar presentation. The symptoms are usually slow in development, although in the cases of metastasis, can occur rather abruptly. An early presentation is *radicular pain*, unilateral or bilateral. Paralysis or paresthesias are also common; usually the paralysis corresponds to the side of the tumor and the paresthesias are contralateral to the paralysis. Often enough, the patient can present with spincter incontinence or urinary retention. Severe back pain should alert the physician, if it is of a long duration. The radicular symptoms usually occur when the tumor arises from a nerve root, although it should always be remembered that pain is not a dominant symptom. However, most of the time, the pain occurs at nighttime and when the patient is in a recumbent position. Because these tumors most commonly occur at the thoracic region, and the pain usually follows the distribution of the intercostal nerve. Occasionally, some patients have been misdiagnosed as having a gallbladder inflammatory process or stomach pain. The tumor can affect one half of the spinal cord and it may produce a Brown-Sequard syndrome, manifested by paralysis on the same side of the tumor with loss of touch on the same side of the tumor and with *contralateral* loss of pain and temperature. The sensory level that is present in Brown-Sequard syndrome is usually two dermatomes below the lesion. If the compression of the tumor compromises the entire spinal cord, it may produce, if it is in the cervical level, quadriplegia and respiratory difficulties.

Sometimes, if the intermedial lateral sympathetic column is involved, the patient can present with hypotension and bradycardia. Paralysis of the sympathetic nerve can produce splanchnic vasoparalysis, and hypotension is produced by massive dilatation of arterioles and veins. In essence, there will be "too much vascular tree" for the same amount of blood, creating a relative hypovolemia. The paralysis of the sympathetic system allows a predominance of the vagus nerve, hence, the bradycardia. When motor paralysis occurs, bilaterally or unilaterally, it is usually spastic. Paroxysmal vascular contractions usually occur at night and the patient complains that his or her legs are jumping. There is a reflex alteration and, although originally the reflexes are depressed, they are usually hyperactive. The patient can present with ankle clonus and Babinski's sign. If the lesion affects the mid- or high-thoracic level, the patient can present with an absence of abdominal cutaneous reflex and cremasteric reflex.

If the pressure of the tumor involves primarily the posterior column, the patient may present with loss of position and vibratory sense. The high cervical tumor and tumor of the foreman magnum may produce headaches, quadriplegia, and sensory loss. With the help of the MRI and myelogram, the tumor can be easily identified. Surgery is the treatment of choice for this type of tumor and can be followed by radiation therapy and/or chemotherapy if the tumor is malignant and/or metastatic.

Bibliography

1. Gray's Anatomy.
2. Bing's Local diagnosis in neurological diseases. The C.V. Mosby Company.
3. Bailey P: Concerning the clinical classification of intracranial tumors. *Arch Neurol Psychiatry* 1921; 5:418.
4. Cairns H: Raised intracranial pressure: Hydrocephalic and vascular factors. *B J.Surg* 1939; 27:275.
5. Courville CB: *Pathology of the central nervous system*, 2nd edition. Mountain View, CA: Pacific Press, 1939.
6. Dickinson W: Parkinsonism secondary to central tumors. *Proc Soc Br Neurol Surg* 1957; 0:69.
7. Baker AB: *Clinical neurology*, 2nd edition. Hoever Medical Division, Harper & Row, 1962.
8. Poggio GF and Mountcastle VB: A study of the functional contribution of the lemniscal and spinothalamic system to somatic sensibility: Central nervous mechanism in pain. *Bull Johns Hopkins Hosp* 1960; 106:266–316.
9. Scheibel ME and Scheibel AB: The spinal arteries. *Acta Radio* 1958; 5:1124–1131.
10. Structural substrates for integrative pattern in the brain stem and reticular cord. In Jasper, H H, et al. *Reticular Formation of the Brain*. Boston: Little, Brown and Company, pp. 31–55.
11. Crosby EC, Humphrey T, and Lauer EW: *Correlative anatomy of the nervous system*. Library of Congress catalog card number: 62–7511. Galt, Ontario: Brett-Macmillan Ltd., 1962.
12. Brodal A: *The reticular formation of the brain stem: anatomical aspect and functional correlation*. Springfield, Illinois: Charles C. Thomas, Publisher, 1958.
13. Mair WGP and Druckman R: The pathology of spinal cord lesions and their relation to the clinical features in protrusion of cervical intervertebral discs (a report of four cases). *Brain* 1953; 76:70–91.

14. Schwartzman RJ and Bogdonoff MD: Proprioception and vibration sensibility discrimination in the absence of the posterior column. *Arch Neurol* 1969; 20:349–353.
15. Barraquer-Ferre L and Barraquer-Bordas L: Trophic disorder of the feet in acquired lesions of the cauda equina. *J Nerv Men Dis* 1952; 116: 902–911.
16 Grinkier RR, Bucy PC, and Sahs AL. In *Neurology*, 5th edition, pp. 344–371. Springfield, Illinois: Charles C. Thomas, Publisher.
17. Mitchell AG: *Anatomy of the autonomic nervous system*. Edinburgh: E. & S. Livingstone, Ltd., 1953.
18. Ruch TC: The urinary bladder. In Fullton, JF, (ed). *A Textbook of Physiology*, 17th edition, pp. 943-949. Philadelphia: W. B. Saunders Co., 1955.
19. Rhoton AL, Jr.: Afferent connections of the facial nerve. *J Comp Neurol* 1968; 133:89–100.
20. Whisler WW and Voris C: Effect of bilateral glossopharyngeal nerve section on blood pressure: a case report. *J Neurosurg* 1965; 23:78–81.
21. Aldrich EM and Baker GS: Injuries of the hypoglossal nerve: Report of ten cases. *Mil Surgeon* 1948; 103:20–25.
22. Hunt JR: On herpetic inflammations of the geniculate ganglion: A new syndrome and its complications. *J Nerv Ment Dis* 1907;34:73–96.
23. Jefferson G and Smalley AA: Progressive facial palsy produced by intratemporal epidermoids. *J Laryng Otol* 1938; 53:417–443.
24. Pearce J 19: The ophthalmological complications of migraines. *J Neurol Sci* 68; 6:73–81.
25. Penfield W and Rasmussen T: *The cerebral cortex of man: A clinical study of localization of function*. New York: The Macmillan Company, 1950.
26. Rodolfo Dassen. Osvaldo Fustinoni. Directores: T. Padilla Y P. Cossio. Sexta Edicion. Libreria "El Ateneo" Editorial. Florida 340 – Cordoba 2099. Buenos Aires. 1955.
27. Raimondi AJ: Hydrocephalus. In *Pediatric Neurosurgery. Theoretic Principles. Art of Surgical Techniques,* pp. 453-489. New York: Springer-Verlag.
28. Benda CE: The Dandy-Walker syndrome or the so called atresia of the foramen of Magendie. *J Neuropathol Exp Neurol* 1954; 13:14-29.
29. Raimondi AJ, Samuelson GS, Yarzagaray L, and Norton T: Atresia of the foramina of Luschka and Magendie. The Dandy-Walker cyst. *J Neurosurg* 1969; 31:202–216.
30. Bucy PC and Kluver H: An anatomical investigation of the temporal lobe in the monkey. *J Comp Neurol* 1955; 103:151–251.
31. Studies of the connection of the fornix system. *J Neurol Neurosurg Psychiatry* 17:75–82.
32. Anatomy of the thalamus. In Schaltenbrand G and Bailey P (eds): *Introduction to Stereotaxis with an Atlas of the Human Brain,* pp. 230–290. Stuttgart: Theime.
33. Brown-Séquard CE: De la transmission croisée des impressions sensitives par la moelle épinière. *C R Soc Biol* (Paris), 1850; 2:33.

34. Bucy PC: Carotid sinus nerve in man. *Arch Int Med* 1936; 58:418.
35. Miller JR: Multiple sclerosis: Diagnosis, diagnostic errors and treatment. *Minnesota Med* 1955; 38:237.
36. Polyak S: Projection of the retina upon the cerebral cortex based upon experiments and monkeys. *A Res Nerv Ment Dis Pro* 1934; 13:535.
37. Morin F and Haddad B: Afferent projections to the cerebellum and the spinal pathways involved. *Am J Physiol* 1934; 172:497–510.
38. *Anatomy of the nervous system.* New York: Appleton-Century-Crofts, 1951.
39. Crosby EC: Relations of brain centers to normal and abnormal eye movements in the horizontal plane. *J Comp Neurol* 1953; 99:437–480.
40. Human thalamus. An anatomical, developmental and pathological study. II. Development of the human thalamic nuclei. *J Comp Neurol* 100:63–97.
41. Engel GL and Aring CD: Hypothalamic attacks with thalamic lesion. *Arch Neurol Psychiatry* (Chicago) 1945; 54:37–50.
42. Brodal A and Walberg F: Ascending fibers in pyramidal tract of cat. *Arch Neurol Psychiat* (Chicago) 1952; 68:755–775.
43. De Jonge B and Crosby EC: The supplementary motor function of the temporal lobe. *Trans Am Neurol Assoc* 1960;171–176.
44. Downman CBB, Woolsey CN, and Lende RA: Auditory I, II and Ep: Qochlear representation, afferent paths and interconnection. *Bull Johns Hopkins Hosp* 1960; 106:127–142.
45. Alterations in response to visual stimuli following lesions of frontal lobe in monkeys. *Arch Neurol Psychiatry* (Chicago) 1939; 41:1153–1165.

Index

Index

The letter *f* following a page number indicates that a figure is being referenced.

A

Abducent eminence, 75–76
Abducent nerve (CN VI). *See* Cranial nerve VI
Acalculia, 18–19
Accessory nuclei, 87, 115
Accommodation, visual, 67, 103
Acoustic nerve, 115, 117
Acoustic neurinomas, 165
Acromegaly, 53, 166
Adamkiewicz, artery of, 134
Adenohypophysis, 52
ADH (antidiuretic hormone), 50
Adie pupils, 105
Adiposogenital dystrophy, 54
Adrenaline, 104
Adversative seizure, 11
Afferent fiber (basal ganglia), 41
Agraphesthesia, 17
Agraphia, 18
Agustia, 119
Akinetic mutism, 13
Ala centralis, 89
Albuminocytolytic dissociation, 62
Alcoholism, 105
Alexia, 18, 22, 26
ALS (amyotrophic lateral sclerosis), 122, 138–139
Altitudinal hemianopsia, 101
Ambiguous nucleus, 86–87, 98*f*, 119
Ammon's horn, 37
Amnesia, 31
Amucia, 22
Amygdala, 30
Amyotrophic lateral sclerosis (ALS), 122, 138–139
Anarthria, 23
Anderson ganglion, 114*f*
Aneurysm, 170
Aneurysmal tumor, 170
Angioblastic meningiomas, 148, 169
Angular gyrus, 5*f*, 7*f*, 15, 18
Anisochoria, 104
Annulus, 149
Anosmia, 99, 164
Ansa hypoglossi, 122, 141
Antebrachial cutaneous nerve, 142
Anterior
  clinoid, 125
  commissure, 36*f*, 59*f*
  fissure (spinal cord), 129
  horn, 130, 133, 138
  lateral column, 131
  orbital convolution, 5
  perforated space, 5*f*
  preolivary syndrome, 87–88
  ramus, 5*f*, 7*f*
  thoracic nerves, 143
  tibial nerve, 149
  white commissure, 56, 129, 131–132
Anterolateral column, 130
Antidiuretic hormone (ADH), 50
Anton's syndrome, 25
Aphasia, 21–23
Apraxia, 17–18
Aqueduct of Fallopius, 113
Aqueduct of Sylvius, 61
Arachnoid membrane, 125–126

Aran-Duchenne atrophy, 139
Arcuate fibers, 84, 91
Area
   17, 25–26
   18, 26
   19, 26
   21, 29
   22, 29
   41, 30
   42, 30
   Broca, 8
   cinerea (fourth ventricle), 85
   compacta (substantia nigra), 66
   medialis (fourth ventricle), 85
   parietal association, 17
   plumiformis (fourth ventricle), 85
   postrema (fourth ventricle), 85
   pretectal, 103f
   reticulata (substantia nigra), 66
Areflexia, 139
Argyll-Robertson pupils, 67, 105f
Arnold-Chiari malformation, 63, 95, 138
Arteriovenous malformation, 31, 169–170
Artery of Adamkiewicz, 134
Ascending frontal convolution, 8
Ascending ramus, 5f, 7f
Asomatognosia, 18
Astereognosis, 17
Asthenia, 92
Astroblastomas, 162
Astrocytomas, 94, 159–160, 174
Ataxia, 17, 47
Ataxia telangiectasias, 169
Atonia, ipsilateral, 92
Auditory nerve (CN VIII). *See* Cranial nerve VIII
Autonomic involuntary nervous system, 158
Autonomic nervous system, 153–158, 153f
   cervical sympathetic system, 156–158
   neurogenic bladder, 136
   parasympathetic bulbosacral outflow, 154f, 155–156
   sympathetic system, 154–155, 154f

B

Babinski-Nagotte syndrome, 88
Babinski's sign, 138–139, 150, 175
Basal ganglia, 39–41
Basilar canal, 83
Basket cells, 91
Basophilic adenomas, 167
Bechterew, nucleus of, 69, 115
Bell's palsy, 113
Benedikt's syndrome, 71, 72f
Betz cells, 8, 16
Biceps reflex, 150
Binasal visual field defect, 101
Bitemporal hemianopsia, 100–101
Blood vessels of brain, tumors of, 169–170
Brachialis cutaneous medialis nerve, 143
Brachial plexus, 129, 141–145
   lesions, 145
Brachium pontis, 74
Brain
   abscess, 171
   basal ganglia, 39–41
   cerebral hemisphere, 1–6
      medial surface, 2–4, 3f–4f
      ventral or inferior surface, 5–6, 5f
   corpus callosum, 2, 3f–4f, 9f, 36f, 55–57, 59f, 60
   frontal lobe. *See* Frontal lobe
   hypothalamus, 49–54, 50f
      hypophysis. *See* Pituitary gland
      mamillary bodies, 51–52
      medial forebrain bundle, 52
   infections, 171–173
   insula, 33–34
   language, speech, and aphasias, 21–23
   limbic system, 35–37, 36f
      hippocampus, 37
   midbrain. *See* Midbrain
   occipital lobe, 2, 25–27
   overview, 1
   parietal lobe. *See* Parietal lobe
   pituitary gland. *See* Pituitary gland
   pons. *See* Pons
   stem, 36f, 59f, 71f, 97f–98f, 133
   temporal lobe, 2, 29–32

thalamus, 43–47, 44f
  epithalamus, 46–47
  habenula, 47
  lateral nuclear mass, 45–46
  tumors, 159–175
    blood vessels, 169–170
    cerebellum, 63, 170
    choroid plexus papillomas, 163
    colloid cysts, 163
    congenital, 168–169
    frontal lobe, 10–11, 13
    gliomas. *See* Gliomas
    intracranial mesenchymal tumors, 163–165
    lipomas, 165
    metastatic, 170
    parietal lobe, 16–17
    schwannomas, 165
    skull tumors, 170–171
    ventricular system. *See* Ventricular system
Broca's aphasia, 22
Broca's area, 8
Brown-Sequard syndrome, 137, 150, 175
Brudzinski sign, 126
Brun's ataxia, 13, 94
Bulbopontine motor nucleus, 87
Bundle of Gowers, 131
Burdach, column of, 131–132, 134

C

Cajal, interstitial nucleus of, 46
Calcaneus medialis branch, 149
Calcar avis, 60
Calcarine fissure, 3
Calcarine sulcus, 4f–5f, 7f
Caloric test, 117
Canal of the ependyma, 130
Capsula externa, 33
Capsula extrema, 33
Carpal tunnel syndrome, 144
Cauda equina, 133, 137
Caudate nucleus, 39
Causalgia, 144
Cavernous sinuses (dura mater), 124
Cavum velum interpositum, 126

Celiac plexus, 121
Celiac semilunar ganglion, 153f
Central cord syndrome, 135
Central lobe (cerebellum), 89
Central nucleus of Perlia, 67
Central nystagmus, 116–117
Central sulcus, 5f, 7f
Central tegmental bundle, 70, 86
Cerebellar hemispheres, 89
  physiopathology, 93–95
Cerebellar peduncles, 90
Cerebellum, 36f, 59f, 89–95, 98f
  cerebellar syndrome, 94–95
  physiopathology of, 92–94
  tumors, 63, 170
Cerebral aqueduct, 4f
Cerebral hemisphere, 1–6
  medial surface, 2–4, 3f–4f
  ventral or inferior surface, 5–6, 5f
Cerebral peduncle, 4f, 36f, 59f, 65, 70, 71f
Cerebrospinal fluid (CSF), 61–62
Cervical enlargement of spinal cord, 129
Cervical ganglia, 156
Cervical plexus, 141–145
Cervical sympathetic nervous system, 156–158
Cervicofacial division of facial nerve, 112
Charcot joint, 138
Check mechanism, 92
Chemical meningitis, 127
Cholesteatomas, 169
Chorda tympani, 111f, 119
Chordomas, 168–169
Choroid plexus, 3f, 37, 59f, 60–61
  fourth ventricle, 85, 126
  papilloma, 63, 163
  third ventricle, 36f, 59f
Chromophobe adenomas, 53, 167
Ciliary ganglion, 107f, 155
Ciliary nerve, 107
Ciliospinal center, 102
Cingulate gyrus, 2–3, 3f, 55
Cingulate sulcus, 3f
Circular sinuses (dura mater), 124
Circumflex nerve, 143
Cisterna magna, 61, 125

Cistern of the pons, 125
Clarke, column of, 92, 130, 132, 134
Clarke, dorsal nucleus of, 78
Claude syndrome, 71, 72f
Clava, 68
Clinging fibers, 91
Clivus, 83, 89
Clostrum, 33
CN. *See* Cranial nerves
Cocaine, 104
Cochlear nerve, 115
Cochlear nucleus, 69, 86
Cognition, 30–31
Collasomarginal sulcus, 4
Collateral fissure, 6, 29
Collateral sulcus, 4f
Colliculus, 36f
Colliculus fascialis, 111
Colloid cysts, 163
Column of Burdach, 131–132, 134
Column of Clarke, 92, 130, 132, 134
Column of Goll, 131–132, 134
Commissural fibers, 91
Common peroneal nerve, 148–149
Communicating hydrocephalus, 63
Complete homonymous hemianopsia, 25
Conaction, 30
Congenital brain tumors, 168–169
Consensual light response, 103
Consensual reflex, 104
Constructional apraxia, 18
Conus medullaris, 129, 134, 136
Convergency, 103
Cordotomies, 135
Corona radiata, 57
Corpora quadrigemina, 65, 70
Corpus album, 36
Corpus callosum (CC), 2, 3f–4f, 9f, 36f, 55–57, 59f, 60
Corpus dentatum, 90
Corpus psalterium, 35
Corpus restiformus, 84
Corpus striatum, 39
Corpus trapezoideum, 74
Cortical lesions, 21
Corticobullar tract, 98f
Corticopontocerebellar tract, 93
Corticospinal pyramidal tract, 10f

Corticospinal tract, 140
COWS mnemonic, 12
Cranial nerves (CN), 97–122, 97f–98f
  CN I (olfactory nerve), 98–99
  CN II (optic nerve), 4f, 99–101, 99f, 103f
  CN III (oculomotor nerve), 67, 69–70, 72f, 98f, 101–105, 101f, 103f
  CN IV (trochlear nerve), 69–70, 98f, 101, 101f, 106
  CN V (trigeminal nerve), 68, 70, 97f, 106–110, 107f–109f
    branches, 107–109, 107f
    inferior maxillary nerve, 108–109
    nuclei, 74–75, 98f
    ophthalmic division, 107, 107f
    superior maxillary nerve, 107f, 108
  CN VI (abducent nerve), 69–70, 101, 101f, 110
    nucleus, 75–76, 98f
  CN VII (facial nerve), 97f–98f, 107f, 109f, 110–114, 111f, 114f
    branches, 112
    nucleus, 76
    paralysis, 113–114
  CN VIII (auditory nerve or vestibular-cochlear nerve), 98f, 114–118
    nucleus, 77–78, 87
  CN IX (glossopharyngeal nerve), 97f–98f, 118–119
    nucleus, 98f
    tympanic branch, 118
  CN X (vagus or pneumogastric nerve), 97f–98f, 119–122
  CN XI (spinal accessory nerve), 97f–98f, 122
  CN XII (hypoglossal nerve), 83, 85, 87, 98f, 122
Craniopharyngioma, 53, 166–167
Craniosacral system, 153
Crescentric lobe, 89
Cruciform sulcus, 70
Cryptococcus neoformans, 171
CSF (cerebrospinal fluid), 61–62
Culmen, 89
Cuneiform nucleus, 66
Cuneus, 3f, 4
Cushing's disease, 167
Cushing's syndrome, 53, 167

Cutaneous muscles of the face, 76
Cystic astrocytomas, 94

D

Dandy-Walker syndrome, 64, 90, 95
Darkschewitsch, nucleus of, 46, 69–70
David's lyra, 35
Deafness, 116
Deep peroneal nerve, 149
Deiters, nucleus of, 69, 77, 115
Déjà vu, 31
Dejerine, semilunar nucleus of, 44
Dejerine-Roussy syndrome, 46
Dentate nucleus, 66, 91
    lesions, 94
Dermatomes, 133–136
Dermoid tumors, 169
Descending virgular tract, 132
Diabetes insipidus, 50
Diabetes mellitus, 105
Diaphragm sella, 125
Digastric lobe, 89
Direct pyramidal tract, 131
Direct spinocerebellar tract, 131
Discogenic syndrome, 149
Dorsal column of Clarke, 92
Dorsal funicular nucleus, 132
Dorsal nerves, 151
Dorsal nucleus of Clarke, 78
Dorsal scapular nerve, 143
Dorsal sulcus (spinal cord), 129
Dorsofunicular white matter, 130
Dorsolateral nucleus (anterior horn), 130
Dorsolateral spinothalamic tract, 68
Dorsomedial sulcus, 129
Dura mater, 123–125
    sinuses of, 123–124
Dysautotopognosis, 18
Dyschoria, 104

E

Edinger-Westphal nucleus, 67, 102, 103f
Efferent fiber (basal ganglia), 41
Efferent thalamic fibers, 17
Encephalon, 1
Endolymphatic fluid, 116

Ependyma, 39, 55, 60
Ependymal cells, 61
Ependymoma, 63, 160, 174
Epidermoid tumors, 169
Epithalamus, 46–47
Equilibrium, 92, 114, 116
Erb-Duchenne palsy, 145
Erythromelalgia, 158
Esinophilic granulomas, 171
Esophageal plexus, 120–121
Exophthalmia, 110
External popliteal nerve, 149
External reticular nucleus of the
    thalamus, 43
Extradural tumors, 174
Extralaminary nucleus, 43
Eye muscles, 101f
    innervation of. *See* Cranial nerves III,
        IV, and VI

F

Facial hemiatrophy, 158
Facial nerve. *See* Cranial nerve VII
Falx cerebelli, 123, 125
Falx cerebri, 123, 125
Fascia dentata, 6, 35
Fasciculi, 12f
Fasciculus arcuatus, 56
Fasciculus cuneatus, 83, 134
Fasciculus gracilis, 83–84, 134
Fasciculus lenticularis, 66
Fasciculus solitarius, 86
Fasciculus thalamicus, 41
Fasciculus tractus solitarius, 119
Fasciola cinerea, 35
Femoral nerve, 147–148
Fiber propria of the cerebellum, 91
Fibers of Rosenthal, 160
Fibrilar astrocytomas, 160
Fibroblastic meningiomas, 148
Filum terminalis, 129
Fimbria, 35, 37
Fissure of Rolando, 1, 8
Fleshing, oval center of, 57
Flexiform ganglion, 119
Flocculonodular system, 92–93
    lesions, 93

Flocculus, 89
Fluent aphasia, 19
Folium, 89
Foniculus incertus, 85
Footdrop, 150
Foramen cecum, 73, 76, 83, 110
Foramen magnum, 83, 123, 129
Foramen of Luschka, 85, 95
Foramen of Magendie, 85, 95
Foramen of Monro, 36, 36*f*, 59*f*, 60–61, 163
Foramen ovale, 108
Forceps anterior, 55
Forceps major, 55
Forceps minor, 55
Forceps posterior, 55
Forel, field of, 41
Forking of the aqueduct, 63
Fornix, 3*f*, 4, 35, 36*f*, 55–56, 59*f*, 61
Foster-Kennedy syndrome, 13, 164
Fourth ventricle, 3*f*, 36*f*, 59*f*, 61, 73, 84–85, 98*f*, 130
 lesions, 93
Friedreich's ataxia, 95, 139
Frohlich's syndrome, 54, 168
Frontal gyrus, 3*f*
Frontal lobe, 2, 7–13, 7*f*, 9*f*–10*f*, 12*f*
 tumor, 10–11, 13
Frontal pole, 5*f*, 7*f*
Fronto-occipital fasciculus, 12*f*
Fusiform gyrus, 29

G

Ganglion
 Andersch, 118
 Anderson, 114*f*
 basal, 39–41
 celiac semilunar, 153*f*
 cervical, 156
 ciliary, 107*f*, 155
 flexiform, 119
 Gasser, 106, 108
 geniculate, 109, 111, 113
 inferior cervical, 156
 inferior mesenteric, 157
 jugular, 118–119
 middle cervical, 156
 oral, 114*f*
 otic, 109*f*, 111*f*
 petrous ganglion of Andersch, 118
 plexiform, 114*f*, 120
 pterygopalatine, 107*f*, 109*f*, 111*f*
 spinal, 153*f*
 submandibular, 111*f*
 superior cervical, 156
 sympathetic, 153*f*
Gastric outlet obstruction, 93
Gelatinous nucleus, 75, 97*f*
General paresis, 172
Geniculate ganglion, 109*f*, 111, 113
Genitocrural nerve, 147
Genitofemoral nerve, 147
Genu (corpus callosum), 55, 57
Gerstman syndrome, 19
Giacomini, tract of, 35, 99
Gigantism, 53, 166
Glioblastoma multiforme, 159, 161, 174
Gliomas, 159–163
 astroblastomas, 162
 astrocytomas, 159–160
 ependymomas, 160
 glioblastoma multiforme, 161
 medulloblastomas, 161–162
 microgliomas, 161
 oligodendrogliomas, 160
 pineal tumors, 162–163
 spongioblastomas, 162
Globus intermedius, 40
Globus pallidum, 39
Globus pallidus, 40
Glossopharyngeal nerve. *See* Cranial nerve IX
Golgi cells, 91
Goll, column of, 131–132, 134
Gowers, bundle of, 131
Gradenigo's syndrome, 110
Gratiolet, optic radiation of, 100
Gray ala, 85, 119
Gray commissure, 130
Gray communicating rami, 153*f*
Gray matter, 67, 102, 130–131, 138–139
Gray pontine nucleus, 74
Gray reticular formation, 87
Greater auricular nerve, 141

Greater occipital nerve, 141
Greater petrosal nerve, 109f, 111f, 114f
Gumma, 172–173
Gyrus fornicatus, 3
Gyrus fusiform, 6
Gyrus longi, 33
Gyrus rectus, 3, 3f, 5
Gyrus supramarginalis, 15

H

Habenula, 47, 52
Hand-Schuller-Christian disease, 170–171
Harris, morbus of, 119
Headache, 171
Hemangioblastomas, 94, 170
Hemianopsia, 25, 53, 100–101
Hemiatrophy of the tongue, 122
Hemifacial spasm, 114
Hemiparesis, 114, 137
Hemiplegia, 114, 137
Herniated disk, 149–150
Herpes zoster virus, 110
Herpetic neuralgia, 139
Heteronymous hemianopsia, 100
Hilus, 86
Hippocampal commissure, 56
Hippocampus, 6, 37
  lesions, 31
Homonomous hemianopsia, 100
Horizontal hemianopsia, 101
Horner's syndrome, 88, 135, 156, 158
Hydrocephalus, 61–64, 93, 163
Hyperacousia, 113
Hyperosmolar diabetic nonketogenic coma, 51
Hyperphagia, 51
Hyperpituitarism, 166
Hypogastric nerve, 148
Hypoglossal nerve (CN XII), 83, 85, 87, 98f, 122
Hypophysis. *See* Pituitary gland
Hypopituitarism, 167
Hypothalamus, 49–54, 50f
  hypophysis. *See* Pituitary gland
  mamillary bodies, 51–52
  medial forebrain bundle, 52

I

Idiokinetic apraxia, 18
Idiokinetic formula, 17
Iliac nerve, 148
Iliohypogastric nerve, 147, 151
Ilioinguinal nerve, 147–148
Induseum grisseum, 35
Inferior
  central nucleus (reticular formation of medulla oblongata), 87
  cerebellar peduncle, 84, 131
  cervical ganglion, 156
  colliculus, 59f, 71f, 98f
  dentary nerve, 109
  frontal gyrus, 3f, 5f, 7f
  frontal sulcus, 5f, 7f
  gluteal nerve, 148
  laryngeal nerve, 121
  longitudinal fasciculus, 27
  longitudinal sinus (dura mater), 124
  medullary velum, 84–85, 90
  mesenteric ganglia, 157
  occipital convolution, 25
  occipitofrontal longitudinal fasciculi, 18
  olivary nucleus, 132
  parietal lobe, 5f, 7f, 18
  petrosal sinuses (dura mater), 124
  sulcus, 29
  tela choroidea, 84–85
  temporal gyrus, 4f–5f, 6, 7f
  temporal sulcus, 4f–5f, 7f
  vestibular nucleus, 133
Inion, 124
Inner periosteum, 123
Insula, 33–34
Insula of Reil, 12, 33–34
Intercostal nerves, 151
Intermediary nerve of Wrisberg, 76
Internal arciform fibers (reticular formation of medulla oblongata), 87
Internal medullary lamina, 44f
Internal popliteal nerve, 149
Interpeduncular space, 65, 102
Interstitial nucleus of Cajal, 46
Interthalamic adhesion, 3f, 36f, 59f
Intracranial mesenchymal tumors, 163–165

Intradural extramedullary tumors, 174
Intralaminary nuclei, 43–44
Intramedullary tumors, 174–175
Intraparietal sulcus, 3, 5f, 7f, 15
Intrapetrosal branches, 112
Intraventricular foramen of Monro, 36f, 59f, 60–61, 163

J

Jackson's syndrome, 88
Jacobson nerve, 114f, 118
Jamais vu, 31
Jugular ganglion, 118–119

K

Kernig's sign, 126
Kinetic center, 17
Kline-Levine syndrome, 51
Kluver-Bucy syndrome, 31
Koniocortex, 16

L

Labyrinthitis, 117
Lamellas, 86
Lamina, 65
Lamina terminalis, 36f, 37, 59f
Lancisi, tract of, 35, 55
Landry, progressive ascendent paralysis of, 139
Language, 21–23
Lateral
   acoustic tubercle, 87
   cutaneous nerve of the thigh, 147
   femoral nerve, 148
   fissure, 1
   geniculate, 4f, 44f, 71f, 99f, 100, 103f
   inferior pontine syndrome, 79–80
   lemniscus, 69–70, 71f, 115
   medullary lamina, 40
   medullary syndrome, 88
   mid–pontine syndrome, 80
   nuclear mass, 45–46
   plantaris nerve, 149
   preopticohabenular tract, 47
   rectus muscle, 76
   sinuses (dura mater), 124
   spinothalamic tract, 132, 138
   sulcus, 4f–5f, 7f
   superior pontine syndrome, 80–81
   tract (medulla oblongata), 83
   ventricles, 60
Lemniscus
   lateral, 69–70, 71f, 115
   medial, 67–68, 98f, 115, 132
Lenticular nucleus, 39–40, 66, 70
Leptomeninges, 126, 165
Lesser petrosal nerve, 111f
Leukocytosis, 171
Lilluputism, 31
Limbic system, 35–37, 36f
Lingual gyrus, 4f
Lingual nerve, 109, 109f, 111f
Lingula, 85, 89
Lipman formula, 17
Lipomas, 165
Lipomatous meningiomas, 148
Lissauer's paralysis, 172
Lissauer's tract, 131, 134
Locus ceruleus, 75, 85, 97f, 106
Logorrhea, 23
Long ciliary nerve, 102
Longitudinal cerebral fissure, 4f
Longitudinal fasciculus, 12f, 70
Long thoracic nerve, 142
Lou Gehrig's disease, 122, 138–139
Lumbar enlargement of spinal cord, 129
Lumbar nerves, 131, 147–151
Lumbego, 149
Lumbosacral cord, 148
Lumbosacral plexus, 147–151
   intercostal nerves, 151
Lunate sulcus, 5f, 7f
Luschka, foramen of, 85, 95

M

Macroglosia, 53
Macular tract, 100
Magendie, foramen of, 85, 95
Magnocellular nuclei, 66
Mammillary bodies, 3f–4f, 36f, 51–52, 59f
Mandibular nerve, 107f, 109f
Marchiaffava-Bignani's disease, 55

Marcus Gunn phenomenon, 105
Masa intermdeia, 61
Mass reflex, 136
Masticatory nerve, 108
Masticatory nucleus, 74–75, 97*f*
Maxillary nerve, 107*f*, 108, 109*f*
Meckel's cavum, 106
Medial
   brachiocutaneous nerve, 142–143
   cerebellar peduncle, 71*f*
   fissure (spinal cord), 129
   forebrain bundle, 52
   geniculate, 4*f*, 44*f*, 71*f*
   gyri, 6
   inferior pontine syndrome, 79
   lemniscus, 67–68, 98*f*, 115, 132
   longitudinal fasciculus, 69, 115
      syndrome, 70
   medullary syndrome, 87–88
   mid–pontine syndrome, 80
   plantar nerve, 149
   sulcus, 85, 129
   superior pontine syndrome, 80
   sural nerve, 149
   vestibulospinal tract, 133
Median nerve, 143–145
Medulla oblongata, 36*f*, 59*f*, 73–74, 83–88, 133, 155
   fourth ventricle, 84–85
   internal structures, 86–87
   reticular formation, 87
   syndromes of, 87–88
      lateral medullary syndrome (Wallenberg syndrome), 88
      medial medullary syndrome (anterior preolivary syndrome), 87–88
      posterior (retro–olivary) syndrome, 88
Medulloblastomas, 94, 159, 161–162, 174
Mendel-Bekhterev reflex, 138
Meniere's syndrome, 117
Meninges, 123–127
   arachnoid membrane, 125–126
   dura mater, 123–125
   infections, 171
   pia mater, 126–127
Meningiomas, 63, 148, 163–164, 169, 173

Meningitis, 126–127
Meningotheliomatous meningiomas, 164
Mentonian nerve, 109
Mesencephalic syndromes, 71
Mesenchymatous meningiomas, 164
Metamorphosis, 31
Metathalamus, 43
Metencephalon, 73
Microgliomas, 161
Midbrain, 65–71
   lateral lemniscus, 69–70, 71*f*, 115
   medial lemniscus, 67–68, 98*f*, 115, 132
   periaqueductal gray matter, 67, 102
   quadrigeminal tubercles, 70–71, 71*f*–72*f*
Middle cervical ganglion, 156
Middle frontal gyrus, 5*f*, 7*f*
Middle primary trunk, 142
Middle temporal gyrus, 5*f*, 7*f*
Midriasis, 102
Millard-Gubler's syndrome, 79
Monro, foramen of, 36, 36*f*, 59*f*, 60–61, 163
Morbus of Harris, 119
Morphine, 104
Motor association field, 17
Motor atonic neurogenic bladder, 136
Motor strip, 9*f*
Movement disorders, 40–41
Multinucleated giant cells, 161
Multiple sclerosis, 94–95, 105, 139
Musculocanateous nerve, 143
Myelomeningocele, 63
Myosis, 102, 104
Myotomes, 133–136

N

Naffziger's sign, 150
Napoleon cap, 47
Nelson's syndrome, 54, 167
Neocerebellum, 89
Neocortex, layers of, 6
Neostriatum, 39
Nerve erigentes, 157
Nerve of Jacobson, 114*f*, 118
Nerves, cranial. *See* Cranial nerves
Nervus hypogastricus, 157

Nervus intermedius, 111
Neuralgia paresthetica, 148
Neurocysticercosis, 171
Neuroepiphysis, 165–166
Neurofibromas, 173
Neurofibromatosis multiplex, 117, 174
Neurogenic bladder, 136–137
Non-inhibited bladder, 136
Nucleus
   accessory, 87, 115
   accumbens septi, 39
   ambiguous, 86–87, 98f, 119
   Bechterew, 69, 115
   bulbopontine motor, 87
   Cajal (interstitial), 46
   caudate, 39
   Clarke, dorsal nucleus of, 78
   cochlear, 69, 86
   cranial nerves. See Cranial nerves
   cuneatus, 68
   cuneiform, 66
   Darkschewitsch, 46, 69–70
   Deiters, 69, 77, 115
   Dejerine, 44
   dentate, 66, 91, 94
   dorsal funicular, 132
   dorsolateral, 130
   Edinger-Westphal, 67, 102, 103f
   emboliformis, 91
   external reticular of the thalamus, 43
   extralaminary, 43
   fastigii, 91
   gelatinous, 75, 97f
   gelatinous of Rolando, 75, 106
   globosus, 91
   gracilis, 68
   gray pontine, 74
   inferior central, 87
   inferior olivary, 132
   inferior vestibular, 133
   intercalatus, 51
   intralaminary, 43–44
   lateralis dorsalis, 45
   lateralis posterior, 45
   lenticular, 39–40, 66, 70
   locus ceruleus, 75
   magnocellular, 66
   masticatory, 74–75, 97f
   olivary, 86
   parataenialis, 45
   perifornical, 52
   Perlia, central nucleus of, 67
   pons. See Pons
   pontis, 78
   posterior salivatory, 86
   pulposus, 137
   red, 66, 71, 72f, 98f, 133
   retrodorsal, 130
   roof, 91
   sacral parasympathetic, 130
   solitary, 121
   superior olivary, 69, 76
   tegmental gray, 66
   vagal, 86
   ventral auditory, 87
   ventralis anterior, 45
   ventrolateral, 130
   ventroposterolateral, 132
   vestibular, 115
   white, 66, 71, 72f, 90
Nylen nystagmus, 117
Nylen two, 117
Nystagmus, 18, 93, 116–117

O

Obturator nerve, 148
Occipital lobe, 2, 25–27
Occipital pole, 3f, 5f, 7f
Occipital sinuses (dura mater), 124
Occipitofrontal fasciculus, 56
Oculomotor nerve (CN III), 67, 69–70, 72f, 98f, 101–105, 101f, 103f
Olfactory bulb, 4f
Olfactory nerve (CN I), 98–99
Olfactory sulcus, 4f
Olfactory tract, 4f
Oligodendrogliomas, 160, 174
Olivary body, 83–84
Olivary nucleus, 86
Open jack-knife phenomenon, 138
Ophthalmic nerve, 107, 107f, 109f
Ophthalmoplegia, 67, 105, 110, 164
Opistotomus, 126
Oppenheim, syndrome of lost hand of, 6
Oppenheim's reflex, 138

Opthalmoplegic migraine, 105
Optic chiasm, 4f, 9f, 36f, 59f, 71f, 99f, 100–101
Optic nerve (CN II), 4f, 99–101, 99f, 103f
Optic radiation of Gratiolet, 100
Optic tract, 4f, 9f, 71f, 99f, 103f
Oral ganglion and nerve branches, 114f
Orbital gyri, 4f, 5
Orbital sulci, 4f
Orientation, 116
Orthosympathetic system, 153
Otic ganglion, 109f, 111f
Oval center of Fleshing, 57

P

Pacchionian granulations, 123
Pain sensation, 17, 68, 130, 132, 137–138
Paleocerebellum, 89
Paleostriatum, 39
Panhypopituitarism, 54, 64
Paquimeninges, 126
Paracentral lobe, 3f, 4
Parahippocampal gyrus, 4f
Paralytic myosis, 104–105
Paraphysis, 60
Paraplegia, traumatic, 137
Parasagittal meningiomas, 148
Parasympathetic bulbosacral outflow, 154f, 155–156
Parasympathetic system, 153, 153f
Parasympathomimetic substances, 156
Parietal association area, 17
Parietal lobe, 2, 5f, 7f, 15–19, 19f
    tumors or lesions, 16–17
Parieto-occipital fissure, 2
Parieto-occipital sulcus, 5f, 7f
Parinaud's syndrome, 71, 72f, 162
Parkinsonianism, 93–94
Parkinson's disease, 66
Parkinson's syndrome, 41
Pars distalis, 52
Pars intermedius, 52
Pars lateralis (substantia nigra), 66
Pars nervosa, 49, 52–53
Pars tuberalis, 52
Parvocellular group, 66
Patellar reflex, 150

Pearly tumors, 169
Peduncular fibers, 90
Pelvic nerve, 135
Pelvic sympathectomy, 137
Perforated substance, 4f
Periaqueductal gray matter, 67, 102
Perifornical nucleus, 52
Peripheral nystagmus, 116–117
Perlia, central nucleus of, 67
Pernicious anemia, 140
Perverted nystagmus, 93
Pervertic nystagmus, 117
Pes peduncularis, 65
Petrous ganglion of Andersch, 118
Phalen's sign, 144
Phantom limb phenomenon, 18
Pharyngeal plexus, 114f
Photomotor center, 67
Photomotor reflex, 103, 103f
Phrenic nerve, 135, 141
Pia mater, 126–127
PICA (posterior inferior cerebellar artery), 90
Pineal body, 36f, 59f
Pineal gland, 3f, 46–47
Pineal recess, 47
Pineal tumors, 162–163
Piriform muscle, 148
Pituitary gland (hypophysis), 36f, 51–54, 59f, 165–168
    stalk, 125
    tumors, 165–168
Pleocytosis, 62
Plexiform ganglion, 114f, 120
Pneumogastric nerve (CN X), 97f–98f, 119–122
Polio, 122, 138
Pons, 3f, 36f, 59f, 71f, 73–78, 84
    auditory nerve nucleus, 77–78
    cranial nerve VI nucleus, 75–76
    cranial nerve V nuclei, 74–75
    facial nerve nucleus, 76
    syndromes of, 79–81
        lateral inferior pontine syndrome, 79–80
        lateral mid-pontine syndrome, 80
        lateral superior pontine syndrome, 80–81

Pons (*Continued*)
    medial inferior pontine syndrome, 79
    medial mid–pontine syndrome, 80
    medial superior pontine syndrome, 80
Pons Varolii, 73
Pontine cistern, 125
Positional paroxysmal nystagmus, 93
Position sense, 138
Postcentral gyrus, 5*f*, 7*f*
Postcentral sulcus, 5*f*, 7*f*
Postclival fissure, 89–90
Posterior
  column, 139–140
  commissure, 36*f*, 56, 59*f*, 103*f*
  horn, 130
  inferior cerebellar artery (PICA), 90
  lateral fissure, 130
  pulmonary plexus, 120
  ramus, 5*f*, 7*f*
  salivatory nucleus, 86
  syndrome, 88
Posterolateral column, 130
Posteromedial column, 130
Praziquantel (Biltricide), 171–172
Precentral fissure, 89
Precentral gyrus, 5*f*, 7*f*
Precentral sulcus, 3*f*, 5*f*, 7*f*, 8
Preclival fissure, 89–90
Precocious puberty, 51
Precuneus, 3*f*, 4, 15
Pretectal area, 103*f*
Primary auditory cortex, 29
Prognathism, 53
Prolactinoma, 53
Proprioception, 138–139
Proprioceptive stimuli, 16–17
Proprioceptive system, 92
Proprioceptive tract, 78
Psammoma bodies, 164
Pterygopalatine ganglion, 107*f*, 109*f*, 111*f*
Pudendal nerve, 135–136
Pulmonary embolism, 16
Pulvinar, 4*f*, 44*f*, 71*f*
Pupillary sphincter, 102–103
Purkinje cells, 91
Putamen, 40
Pyramidal tract, 9*f*, 98*f*, 137–139
Pyramis, 89

Q

Quadrantianopsia, 101
Quadrate lobe, 15
Quadrigeminal plate, 65
Quadrigeminal tubercles, 70–71, 71*f*–72*f*

R

Radial nerve, 143
Rathke's pouch, 52
Raynaud's disease, 158
Raynaud's phenomenon, 158
Rectangular lobe, 4
Recurrent laryngeal nerve, 121
Red nucleus, 66, 71, 72*f*, 98*f*, 133
Reil, insula of, 12, 33–34
Restiform bodies, 84
Reticular formation, 74, 87, 133
Reticular system, 66
Reticulospinal tract, 133, 135
Retrodorsal nucleus (anterior horn), 130
Retro-olivary syndrome, 88
Retroreflex bundle of Meynert, 36
Rheumatoid arthritis, 158
Rinne test, 116
Rochon-Duvigneaud syndrome, 110
Rolando, fissure of, 1, 8
Rolando, gelatinous nucleus of, 75, 106
Romberg sign, 94, 138
Roof, nuclei of, 91
Rosenthal, fibers of, 160
Rose's field, 37
Rossolimo's sign, 138
Rostrum, 55
Rubrospinal tract, 66, 132–133

S

Sacculus, 116
Sacral nerves, 148
Sacral parasympathetic nuclei, 130
Sacral plexus, 148
Saphenous nerve, 148
Sarcomas, 165
"Saturday night palsy of the alcoholic," 143
Scalenous anticus muscle syndrome, 145
Schaffer's reflex, 138
Schmid syndrome, 88

Schnider's syndrome, 135
Schultze, cells of, 98
Schultze, tract of, 132
Schwannomas, 165
Sciatica, 149
Sciatic nerve, 148–149
Scleroderma, 158
Semicircular canal, 116
Semilunar nucleus of Dejerine, 44
Sensory aphasia, 18, 22
Sensory atonic neurogenic bladder, 137
Sensory speech, 22
Septum pellucidum, 36f, 55–56, 59f
Sheehan's syndrome, 54, 167
Shoulder-hand syndrome, 144
Sigmoid gyrus, 14
Sign
   Babinski, 138–139, 150, 175
   Brudzinski, 126
   Kernig, 126
   Naffziger, 150
   Phalen, 144
   Romberg, 94, 138
   Rossolimo, 138
   Tinnel, 144
Skull, tumors of, 170–171
Slow pursuant system, 18
Soft palate, 121
Solitary nucleus, 121
Somesthetic cortex, 16
Spasmodic myosis, 104
Speech, mechanism of, 21–23
Spinal accessory nerve (CN XI), 97f–98f, 122
Spinal cord, 129–140
   commissure, 130
   compression, 174
   diseases, 139–140
   myotomes and dermatomes, 133–136
   neurogenic bladder, 136–137
   trauma, 135–139
   tumors, 173–175
      extradural tumors, 174
      intradural extramedullary tumors, 174
      intramedullary tumors, 174–175
Spinal ganglion, 153f
Spinal nerves, 131, 153f
Spinal reflex, 136

Spinal shock, 137
Spinocerebellar tract, 130
Spino-olivary tract, 132
Spinotectal tract, 132
Spinothalamic tract, 132
Splanchnic nerve, 153f
Splenium, 2, 4f, 55
Spongioblastomas, 162
Status epilepticus, 31
Straight gyrus, 4f
Straight sinus (dura mater), 124
Stratum lemnisci, 71
Stratum zonale, 70–71
Stria acoustica, 85, 115
Stria longitudinalis, 55
Stria medullaris, 85
Stria terminalis, 39
Sturge-Weber disease, 169
Stylomastoid foramen, 113
Subarachnoid space, 61
Subdural space, 125
Submandibular ganglion, 111f
Subscapular nerve, 143
Substantia gelatinosa, 106, 132
Substantia nigra, 4f, 65–66, 72f
Sulcus, 2–3
Superficial peroneal nerve, 149
Superficial petrosal nerve, 118
Superior
   cerebellar peduncle, 71, 72f
   cervical ganglion, 156
   colliculus, 4f, 59f, 71f, 132
   fovea, 85
   frontal gyrus, 3f, 5f, 7f
   frontal sulcus, 5f, 7f
   gluteal nerve, 148
   hypogastric plexus, 137
   laryngeal nerve, 120–121
   longitudinal fasciculus, 56
   longitudinal sinuses (dura mater), 124
   medullary velum, 84, 90
   olivary nucleus, 69, 76
   operculum, 1
   parietal lobe, 5f, 7f
   petrosal sinuses (dura mater), 124
   primary trunk, 142
   temporal gyrus, 5f, 7f
   temporal sulcus, 5f, 7f

Supraclavicular nerve, 141
Supramarginal gyrus, 5f, 7f, 18
Supraolivary fossa, 113
Supraopticohypophyseal system, 50
Supraorbital sulcus, 3
Suprascapular nerve, 143
Swallowing, difficulty in, 119
Sylvian fissure, 1
Sylvian sulcus, 4f
Sylvius, aqueduct of, 102
Sympathetic ganglion, 153f
Sympathetic nervous system, 153, 153f, 154–155, 154f
Syphilis, 67, 104–105, 138, 172
Syphilitic arteritis, 172
Syringomyelia, 138–139

T

Tabes dorsalis, 136, 138, 173
Tactile sensation, 17, 68, 132, 137–138
Tapeworm infestation of the brain, 171
Tapia syndrome, 88
Tardy ulnar nerve palsy, 144
Tegmental gray nuclei, 66
Tegmentum, 65
Tela choroidea, 60–61
Temperature sensation, 17, 68, 130, 132, 137–138
Temporal hemianopsia, 53
Temporal lobe, 2, 29–32
Temporal pole, 4f–5f, 7f
Temporofacial branch of facial nerve, 112
Tentorium, 89, 123, 125
Teratomas, 169
Teres eminence, 85
Thalamus, 3f–4f, 36f, 43–47, 44f, 59f, 132
  epithalamus, 46–47
  habenula, 47
  lateral nuclear mass, 45–46
Thermoanesthesia, 138
Third nerve palsy, 105
Third ventricle, 44f, 60–61
  colloid cyst, 64
Thoracic nerves, 131

Thoracodorsalis nerve, 143
Thoracolumbar outflow, 154
Thoracolumbar system, 153
Thromboembolism, 16
Tibial nerve, 148–149
Tinnel sign, 144
Tolosa Hunt, 110
Tonsils, 90
Tract of Burdach, 131–132, 134
Tract of Giacomini, 35, 99
Tract of Goll, 131–132, 134
Tract of Lancisi, 35, 55
Tract of Lissauer, 131, 134
Tract of Schultze, 132
Transverse cervical nerve, 141
Transverse fissure, 60
Transverse occipital sulcus, 5f, 7f
Transverse sinus (dura mater), 125
Trapezoid body, 69, 74, 77
Treponema pallidum, 172
Triceps reflex, 150
Trigeminal nerve. *See* Cranial nerve V
Trigeminal neuralgia, 110
Trigone, 60
Trigonum acousticus, 85
Trigonum of the hypoglossal nerve, 85
Trigonum of the vagus nerve, 85
Trochlear nerve (CN IV), 69–70, 98f, 101, 101f, 106
Tuber, 89
Tuber cinereum, 50
Tuberculomas, 172
Tuberculous meningitis, 127
Tumors. *See also* names of specific tumors (e.g., Astrocytomas, Meningiomas)
  aneurysmal, 170
  brain. *See* Brain tumors
  dermoid, 169
  epidermoid, 169
  extradural, 174
  gliomas. *See* Gliomas
  intradural extramedullary, 174
  intramedullary, 174–175
  pearly, 169
  pituitary gland (hypophysis), 165–168
  spinal cord, 173–175

Turcular herophili, 124
Turk Maynert, 74

U

Ulnar nerve, 142, 144–145
Uncal herniation, 104
Uncinate fasciculus, 12*f*, 133
Uncus, 4*f*, 6
Urinary bladder, 157
    innervation of, 135–136
Utricle, 116
Uvula, 89–90

V

Vagal nucleus, 86
Vagus nerve (CN X), 119–122, 979*f*–98*f*
Vegetative nervous system, 158
Velum interpositum, 60
Ventral auditory nucleus, 87
Ventral corticospinal tract, 98*f*
Ventral horn cells, 133
Ventral spinothalamic tract, 68, 132
Ventral white commissure, 132
Ventricular funiculus, 133
Ventricular system, 59–64
    aqueduct of Sylvius and fourth ventricle, 61
    cerebrospinal fluid (CSF), 61–62
    hydrocephalus, 62–64
    lateral ventricles, 60
    overview, 59, 59*f*
    third ventricle, 60–61
Ventriculoperitoneal shunt, 161
Ventrolateral nucleus (anterior horn), 130
Ventrolateral vestibulospinal tract, 133

Ventroposterolateral nucleus (VPL) of the thalamus, 132
Vermis, 89, 93
    lesions, 93–94
Vestibular–cochlear nerve. *See* Cranial nerve VIII
Vestibular nerve, 115, 118
Vestibular nucleus, 115
Vestibular system, 116, 133
Vestibulo-spinal fibers, 77
Vibratory sense, 138
Viral meningitis, 127
Vision, 25
Visual field defects, 99*f*
Von Hippel-Lindau disease, 169
Von Recklinghausen's disease, 165
VPL (ventroposterolateral nucleus) of the thalamus, 132

W

Wallenberg syndrome, 88
Weber's syndrome, 71, 72*f*
Weber test, 116
Werdnig-Hoffmann disease, 138
Wernicke-Korsakoff syndrome, 51
Wernicke's aphasia, 19, 23
White commissure, 130
White communicating rami, 153*f*, 154
White internal ala, 85, 122
White matter, 130–131, 139
White nucleus, 66, 71, 72*f*, 90
Wrisberg, intermediary nerve of, 76

Z

Zona incerta, 41

987

Cranial Nerves

I Olfactory
II Optic
III Oculomotor
IV Trochlear
V Trigeminal
VI Abducent
VII Facial
VIII Vestibular-Cochlear
IX Glossopharyngeal
X Vagus / Pneumogastric
XI Spinal Accessory (ALS, Polio)
XII Hypoglossal

LINARDAKIS: DIGGING UP
THE BONES: NEUROSCIENCE